Françoise Malby-Anthony

with Kate Sidley

The Elephants of Thula Thula

MACMILLAN

First published 2022 by Macmillan
an imprint of Pan Macmillan
The Smithson, 6 Briset Street, London EC1M 5NR
EU representative: Macmillan Publishers Ireland Ltd, 1st Floor,
The Liffey Trust Centre, 117–126 Sheriff Street Upper,
Dublin 1, D01 YC43
Associated companies throughout the world
www.panmacmillan.com

ISBN 978-1-5290-8769-7

1 3 5 7 9 8 6 4 2

A CIP catalogue record for this book is available from the British Library.

Typeset by Palimpsest Book Production Ltd, Falkirk, Stirlingshire
Printed and bound by CPI Group (UK) Ltd, Croydon, CR0 4YY

Visit **www.panmacmillan.com** to read more about all our books
and to buy them. You will also find features, author interviews and
news of any author events, and you can sign up for e-newsletters
so that you're always first to hear about our new releases.

The Elephants of
Thula Thula

Also by Françoise Malby-Anthony

An Elephant in My Kitchen

To all animal lovers around the world that share the same passion and vision for wildlife conservation, and to my amazing Thula Thula Wild Team who stands by me, always, in my wild African adventures.

Contents

1

An Elephant in My Garden

What was Frankie doing, standing right in front of my gate?

I kept an eye on her – you always had to with Frankie; she was unpredictable. You never quite knew what our feisty elephant matriarch would do. She turned to face the house and took a few steps, tall and proud. It almost looked as if she were walking inside my garden, I thought with a chill. But that was impossible. There were five electric wires laid on the ground across the open entrance carrying 8,000 volts of electricity to keep the animals out. There was no way Frankie could have breached the boundary.

I walked closer for a better view. Was I dreaming? Frankie was indeed in my garden! Somehow, she had stepped over the wires – was the electricity not working? – and she was walking up to my house. Strolling with confidence into this unauthorized land.

My first panicked thought was, '*Where are my dogs?*' I looked around wildly. An encounter between seven barking dogs and a massive elephant would certainly end in disaster.

I felt hysterical, but kept my trembling voice low, 'Here Tina . . . Lucy, Miley, come . . .' They can be disobedient little chaps, but this time they caught the urgency in my voice and followed me up to the house, all of us moving quickly and quietly. They ran in and I shut the door and leaned against it, shaking with fear and adrenaline.

'Come, come, my doggies, ssshhh . . .' I said, gathering them to me and stroking them to keep them quiet. My little Gypsy shivered against me. Even the naughty yappy ones, the French poodles, Alex and Shani, were on their best behaviour – most dogs have a sensible respect for the elephants. As for Frankie, she had no love for dogs. In fact, she hated them. Gin is lucky to be alive after he foolishly charged Frankie some years ago, and Frankie gave him as good as she got, charging him back. His feet barely touched the ground as he'd fled.

Through the panes of glass in the flimsy wooden framed door, we all watched Frankie.

I couldn't believe what I was seeing! It was July 2018 and Frankie had never come into my garden in all these years, and there she was walking around the place in a very calm, confident way. She wasn't aggressive, or stressed. It was as if she were taking a little stroll around her own home.

She proceeded slowly towards me, one huge foot after the other, until she was no more than five or six metres from my door. Frankie could have knocked that door down with a little flick of her trunk if she wanted to.

Frankie and I have a special – if rather complicated – relationship. Twenty years earlier, soon after she arrived at Thula Thula with the original herd, she nearly killed me and my husband Lawrence when we surprised her on a noisy quad bike. I have never forgotten the terror of seeing her hurtle towards us, ears back, eyes blazing, her furious trumpeting splitting the air. I thought my last day had come.

We lived to tell the tale, and Lawrence named her after me, saying that she shared my feisty French temperament. Frankie had been more than feisty in the beginning. She had a temper and an unpredictable streak that made us all a bit nervous, especially when we had guests on game drives. As she aged, she became calmer and more confident, but she was never one to be taken lightly. The two of us had a healthy mutual respect.

And now here we were, the two Frankies, separated by a few metres of lawn and a few bits of wood and glass.

The elephant looked at us and seemed to hesitate, and for a scary moment, I was sure she was going to come in. Then she turned away from the door and strolled in the direction of the swimming pool.

Behind her, the rest of the herd were gathered at the fence, all twenty-eight of them, from Mandla, our biggest bull, to little Themba, tagging along behind his mother Nandi. The elephants were as surprised as I was to see their matriarch taking a walk around an area that they all knew full well was out of bounds. Nana, our gentle and dignified old lady, Frankie's predecessor as matriarch, must have been shocked at this uncouth behaviour.

As the head of the herd, the matriarch shows leadership and demonstrates correct elephant behaviour. Frankie's job was to set an example to the other elephants and to discipline anyone who stepped out of line. And yet here she was, blatantly flouting the rules herself.

Brendan, her son, decided, 'Well, if Mum can do it . . .'. and walked up to the gate. Was I going to have two elephants in my garden? Or even the whole herd, following their leader?

As Brendan stepped onto the wires we heard a crackling, snapping electrical sound followed by his furious trumpeting scream. The wires were definitely working, and he had got a big shock. Brendan backed off.

Frankie continued her tour of the premises for almost an hour, unhurried and curious. She took in the view over the grasslands down towards the dam. Paused to admire the delicate pink flowers on the kapokboom tree. Rested a moment in the shade of the enormous sycamore fig. She raised her trunk and sniffed the light breeze that brings relief on hot summer days. It was as if she was taking stock of it all with a view to purchase, 'Hmm, it's a nice house; it might suit me.'

I was beginning to wonder if she had indeed taken up residence, and I was going to be stuck in my house with an elephant in my garden for ever, when she turned towards the gate.

All eyes were on Frankie.

The dogs and I watched through the door, and the elephants watched from outside the fence. What was she going to do? Frankie made her slow and deliberate way to the exit. It was a wonder that she'd got in without an electric shock. How was she going to get out? My great fear was that she would get a jolt, and I'd have a furious elephant on my doorstep.

She raised one massive foot and placed it carefully between the wires. By now, the rest of the elephants were in quite a state, shuffling about and looking on anxiously. Some were trumpeting their concern. Others were pointing to the ground with their trunks, almost as if they were saying, 'Be careful . . . Look there's a wire . . . Mind, there's another one . . . Watch your step Frankie.'

Frankie remained calm, raising the next foot, and then the next, placing each one delicately on the ground, avoiding the wires with an acrobatic elegance you would never imagine was possible from a four-ton elephant.

As she cleared the last wire, the herd welcomed her back with their trunks held high in triumph and celebration. There was rumbling, and someone gave a short blast of a trumpet. You didn't need to understand Elephant to know that they were saying, 'You made it Frankie! You're back! Well done!'

Frankie turned her great head to me, as I cowered behind the flimsy door and clutched my dogs. Her eyes met mine and she gave a small toss of her head, as if to say, 'Who's the matriarch now, Madame? I know you thought it was you, but who's the boss really?'

The following evening, I was home alone with the dogs. They seemed a bit edgy, not settling happily into the sofa as usual. Now, when you live in the bush and your dogs behave in an unusual way, it's a good idea to check it out because there is often something amiss. Usually, it is just a monkey in the trees, but it could be something more worrying, like a snake at the door. I moved little Gypsy from my lap and got up to look out of the window.

It was night and I didn't have my glasses on, but I saw nothing out of the ordinary. Just the African flame tree that delighted me with its magnificent red flowers in spring, a big dark shape against the starry sky.

It was a very still night, not a breath of wind, but strangely, the tree was moving as if ruffled by a breeze. I felt a prickle of fear and recognition – that big dark shape wasn't a tree. It was an elephant.

I sent a message to the WhatsApp group of rangers and key staff:

'Shit! Frankie is back in my garden!'

Again, she was very close. Just metres away on the other side of the door, down the two little steps that separated my house from the lawn.

I sent another message:

'Can elephants climb steps?'

Ranger Promise replied with a string of emojis of pairs of round eyes.

Not very helpful! In fact, nobody knew what to do. You can't just shoo a four-ton elephant out of your garden as you can an antelope or a baboon or any of the other creatures who occasionally invade my private space and create havoc with my dogs.

Quite apart from the breach of the perimeter, which Frankie knew full well was prohibited, this was very unusual behaviour. A matriarch should never leave the herd at night. Frankie was alone, so where were the other elephants? They might have been close by but I could not hear or see them in the dark.

Frankie stayed in the garden all night. She had obviously taken a liking to my place – I hoped she didn't intend to move in!

'You have more than 4,000 hectares to wander around, Frankie,' I muttered under my breath.' 'Why do you have to come into my little garden?'

She pottered about, devouring my newly planted indigenous gerbera daisies, and enjoying the night air. She was still for quite a while and I thought perhaps she was having a quick nap standing up, as elephants do. And as dawn broke over the bush, she left as she had arrived, completely noiselessly, stepping delicately over the electric wires and heading down the road to join the family.

A few days later, when I was in the UK for the launch of my first book, Frankie came back into my little garden. The staff were in a panic, and Lynda, manager of admin and accounts, called Andrew, one of the rangers, to help. Quite what she expected him to do, I don't know. Still, Andrew decided to have a little chat with Frankie. With the rest of the staff watching from the safety of my house, he spoke quietly, trying to talk her into leaving.' 'Go on Frankie, time to go home. Come on . . .'

The herd was at the gate watching this scene with intense interest. Gobisa, Frankie's companion, Mabula and Ilanga and Brendan, her sons, and Marula, her daughter, were particularly fascinated by the sight of their mother gate-crashing again.

Andrew has a special bond with all the elephants. Somehow, his quiet words seem to get through to them. But Frankie was in a bad mood that day and was having none of his calm talk and sensible suggestions. She charged, running at him, ears flapping. He ran up the three steps of my veranda and ducked into my house just in time.

After that, they decided to leave her alone doing her round of my little garden, and she eventually left as she came in.

As always, when presented with a mystery or an issue concerning the herd, I consulted the rangers and Christiaan, the conservation manager. No one had an explanation for her strange behaviour. Elephants don't usually come so close to the places where we humans and dogs and cars congregate, unless they're passing by.

'These visits are getting to be a problem,' I said to Vusi, the farm manager who is in charge of all the difficult things like fences and water supply. And now, a breaking-and-entering elephant. 'We can't be having Frankie pottering around the garden whenever she feels like it!'

'It's dangerous,' he agreed. 'For the staff, too. And even the guests.'

My house is part of the compound we call main house. The reception and the admin office are in the original farm house, with its beautiful old Cape Dutch style front. Clustered around it are a number of cottages and outbuildings, including my house and office, and accommodation for some of the office staff. If Frankie could come into my garden, she could pay them a visit too.

'Maybe she's big enough to just step over most of the wires. I'll get security to put an extra wire on the ground,' said Vusi. 'That should discourage her.'

'Thanks Vusi. It will be nice to go to sleep at night without wondering if there's an elephant in my garden.'

Vusi arranged the extra wire, but Frankie came back one more time.

One morning, not long after Frankie's night-time visit, I heard a commotion outside the office. Portia, the marketing assistant, was waving her arms expressively, and regaling the office staff with some tall tale.

She beckoned me over, 'Françoise, you'll never believe what just happened!'

'Well, tell me!' I said. Now, Portia is by nature very dramatic, so I was expecting something like a snake in a cupboard or a frog in her shoe. There's no shortage of such incidents in the bush.

She continued, eyes wide. 'I was in my swimsuit, going to the pool for my early morning swim, and I was in the middle of the lawn when from the corner of my eye I saw something. A big grey rock. For some reason, I didn't register that there isn't a big rock right there on the lawn. I carried on to the pool. Then I heard Andrew's voice, in a shouting whisper, "Portia, Portia, go back . . . Frankie's in the garden!" I turned my head and saw that the big grey rock was Frankie's butt!'

The whole office burst out laughing.

'It's not funny. She was right there next to me! I turned and ran for my life and banged on Swazi's door.'

Swazi, the reservation officer, took up the tale.

'There was Portia wrapped in a towel, with her eyes wide, saying "Swazi, Swazi, let me in. Frankie's in the garden!" She came into my room and shut the door. I ran to the other side and opened the blinds and there was Frankie, right in front of us. This huge elephant, right outside the window. Portia and I were shaking with fear!'

The rest of the office laughed and shrieked and shivered.

The office staff still talk about the time Portia nearly collided with Frankie's butt on her way to the pool. The funny thing is that when we talk about that encounter, Portia remembers it as something quite precious: 'Frankie was so beautiful and so close. It was an unforgettable experience that I'll cherish forever.'

That's Frankie for you – she really is a very special elephant.

On her way out of the property that day, Frankie got a sharp electric shock from the wires. And that was the end of her house calls.

To this day, those visits from Frankie in my garden remain a total mystery. We can only use our human minds and eyes and hearts to imagine a reason. Did she want to visit the kitchen where baby Tom had been rescued years before? Was there something else she was trying to communicate to us?

So many questions we couldn't answer.

2

The Land Before We Lived Here

Before Africa was divided up into countries, provinces, towns and farms. Before the railways and highways and fences and border posts. Before the shopping malls and office blocks and suburbs, elephants moved freely around the continent.

As a migratory species they might travel as much as a hundred kilometres to find food or water, to get away from danger, or to find a more hospitable environment. Some of the colonial roads and railways were even built on the elephants' migratory paths, like the one that crosses the Drakensberg mountains.

Elephants in our province of KwaZulu-Natal, in the northeast of the country, might have travelled as far as Mozambique, allowing overpopulated herds to spread out into new terrain. These days elephants are confined to smaller pockets of wilderness as their habitat is destroyed or encroached by humans.

The more of us humans there are, and the more land we occupy, the more conflict there will be between us and elephants. Our own beautiful herd only came to us because they were breaking out of the game reserve they were living on to eat the neighbouring farmers' crops.

This situation, or a version of it, plays out all over the world. In India, the space for elephants is shrinking, and elephants are chased and even shot by farmers and villagers. We forget that these magnificent beasts were here long before us.

It's no secret that elephants are very large and they eat a lot – an elephant might eat as much as two hundred kilograms a day. That's a lot of trees and bush, so they need a good, big space. If confined to a small area, they can destroy their habitat.

When Lawrence and I bought Windy Ridge, it was 1,500 hectares. As 400 of those hectares were on the other side of the public road, they couldn't be part of the regulatory fenced area where we could keep the wildlife. Local people grazed their cattle on that land. We were left with a little sanctuary of just over 1,000 hectares, which we called Thula Thula. In Zulu, *thula* means quiet, and is often said in hushed tones. A mother might whisper, '*thula, thula*' to comfort a child to sleep. This was the peace and tranquillity we wanted to give animals and humans, in the land that had been hunted on for centuries.

In August 1999, we got our first seven elephants. They had the reputation of being 'problem' elephants and would have been culled if we had not taken them in. Lawrence knew that if we wanted to add to our herd, either through breeding or by rescuing more elephants, we would need more land. As well as the practical, common sense need for space, there's a regulatory one – the elephant management regulations of the Department of Environmental Affairs require us to have a certain number of hectares of land per elephant.

Lawrence was a man with great vision and big plans. His dream was to create a massive conservancy in Zululand, incorporating our land and other small farms and community land into one great big game park, stretching all the way to the far north of the province.

'Imagine it, Frankie,' he would say to me, gesturing over our property and beyond to the horizon, his face bright with excitement. 'All that beautiful bush, the animals. One big, safe, well-managed reserve. All the way up to Umfolozi.'

It was a grand idea, but we had no spare cash in those days. We were just starting out. No one knew us. We couldn't raise donations for land expansion. But Lawrence was undeterred. He spent hours and days trying to rally people in support of his vision. There were endless exhausting meetings with community leaders. His commitment to his dream, and to the welfare of animals, never wavered.

He had some success. In 2008, ten years after we bought Thula Thula, we expanded into 1,000 hectares of land which belonged to the National Parks Board but had been allocated for community use. This area, Fundimvelo, had no water, so it really wasn't suitable for cattle. What little wildlife there was on the community land struggled to survive, and animals often came through or over the fence to find water and better grassland at Thula Thula.

Lawrence approached the *amakhosi*, the local traditional leaders. They are largely symbolic figureheads, most of whom have little political power, but they play an important role in the lives of rural people, negotiating, advising and helping to resolve disputes. Our relationships with the *amakhosi* are extremely important to us, and over the years we have developed great trust and mutual respect.

Lawrence's proposal was to create a joint conservation project on Fundimvelo. We would run it as part of Thula Thula and develop its infrastructure. They didn't hesitate. We dropped the fence between the two properties. Our beautiful family of elephants had more space to roam, and the animals on the community land had access to our water and grasslands.

Lawrence and Vusi and the team built a large dam on what had been community land. It was a favourite spot for our elephants, and of course our hippo family. After Lawrence died in 2012, we named the dam Mkhulu Dam, in memory of him. *Mkhulu* is the Zulu word for grandfather, and was our staff's affectionate name for Lawrence. His ashes were scattered at that beautiful, tranquil place.

When Lawrence died, he had been in discussion with the *amakhosi* about an additional piece of land, bordering Fundimvelo, perfect for the expansion of Thula Thula. I took over the complex negotiations, navigating Zulu customs and land rights, government regulations, and conservation issues – all of it conducted in English and Zulu. It was quite a learning curve for a French-speaker who just wanted to get on with making her little piece of earth a sanctuary for wildlife. Now, seven years later, in 2019, we had yet to finalize the negotiation, but I remained determined to move forward, towards Lawrence's dream.

I often visit Mkhulu Dam of an evening, as the sky turns pink and gold and the hippos grunt and snort, making rings and ripples in the sunset reflected in the water. It is hard to believe that this dam, alive with birds and insects and animals, dried to hard, cracked mud in the terrible drought that began in 2013 and continued for three long years. Our hippos, Romeo and Juliet, and the crocodile Gucci, left Mkhulu Dam, led by their incredible survival instinct to the far side of the reserve where one little dam still had a bit of water.

The rains came, as they always do, and brought life quickly back to normal. The dam filled up and our hippos, crocodiles and other wildlife returned to their home. The whole reserve felt alive and bountiful, the plants and animals thriving. I marvelled again at nature's resilience, and the example it offers for us humans. If you can just hang on long enough, the rains will surely come.

Mkhulu Dam is a favourite place to visit on a game drive. Guests go on two drives a day, one in the early morning, and the other in the late afternoon. The rangers take them out into the bush, tracking and observing the animals – the elephants in particular, of course – and sharing fascinating facts and exciting tales. The guests always leave saying they have learned so much.

These drives really are the highlight of a trip to the bush. There's a great sense of freedom and excitement as you set off in an open vehicle, never knowing exactly what the day will bring. No two drives are the same. But there will always be something thrilling, or amusing, or surprising, something that will touch your heart or soul, make a perfect memory or photograph.

Every drive ends with a picnic stop somewhere – hot coffee

for the morning drive, cocktails for sundowners. Mkhulu Dam is a good spot for such a stop, particularly if the elephants have come down to the dam for a drink, or to cool off in the water. The ideal elephant sighting is for the animals to be undisturbed by human presence. 'I want to see them as they are, doing what they would be doing even if I wasn't there,' says ranger Victor. 'This is the experience I want for our guests too, the feeling of simply being here in this moment with these elephants, right now, in this beautiful place.'

Occasionally, I will join the rangers, and end a busy week watching the sun go down over the bush.

One day in particular stays in my memory. It was almost as if the elephants knew it was Friday afternoon, time to kick back, relax, hang out with friends. The whole family had taken a stroll down to their beloved Mkhulu Dam. As had we humans – well, we'd taken a drive, rather than a stroll, and parked alongside the dam.

It was a gorgeous, soft evening. The hippos surveyed us from the water, just their eyes, ears and nostrils visible. A mother francolin cackled at her chicks to follow her into the long grass. A pied kingfisher hovered above the dam on a blur of beating wings and plunged into the water, emerging with a tiny, wriggling tilapia fish.

The elephants were at their most charming and photogenic. Marula took a gentle promenade along the water's edge. The youngsters were in good spirits, splish-sploshing about, kicking up the mud with their feet.

Our guest had come all the way from America to fulfil a

long-held dream of seeing Thula Thula and meeting the elephants she had fallen in love with from afar. A keen amateur photographer, she couldn't stop taking pictures – I think her fingers must have been sore from pressing that shutter. And they were indeed very beautiful.

Like proud parents, we surveyed our happy herd and discussed them. The rangers at Thula Thula know each elephant, their name and history, who their parents are, which are their siblings. They know their personalities and moods. They can tell when they are relaxed, when they want to be given some space, if they are feeling playful, or just curious.

'Haven't the little ones grown?' I said to Khaya, our youngest ranger. 'I can't believe how big Tom is.'

'She wouldn't fit into your kitchen now Madame,' he said. 'Not even through the door!'

As a newborn baby, Tom had wandered away from the herd and under the electric fence, and found her way to my house – and subsequently became the star of my first book *An Elephant in My Kitchen*. Seven years later, she is a beautiful, feisty young lady who likes to practice flapping her ears and looking fierce to try to give you a scare. Tom also loves the sound of her own voice – she has a distinctive trumpeting sound – and is quite a noisy little elephant.

'And what about my boy, isn't he handsome?' said Vusi, pointing proudly to his own namesake, little Vusi, just a year older than Tom.

'Aw, of course,' I said indulgently. 'Your Vusi is a very beautiful little elephant.'

'Maybe he takes after the other side of the family,' said

Khaya, jokingly. The rangers love each other like brothers, but they do like to tease each other.

'You can only see the scar from the snare on his face if you really look. It has healed so well,' said Vusi, his binoculars up to his eyes.

At just a week old, the little elephant had been caught in a poacher's snare. It wrapped around his mouth and he couldn't suckle. His mother, Marula, found Vusi, who was doing his rounds on the reserve, and gently pushed the little elephant towards him. Human Vusi got the message and called our vet, who managed to remove the snare.

Vusi is a lucky elephant. He would have starved if we hadn't helped him.

'Hello Bafana,' said Khaya, pointing out his own favourite. 'Ah, look at him, what a fine big guy he is. And there's Kink. On the edge of the herd as usual. He's a shy one.'

Kink – Nandi's youngest and Nana's grandson – is always easy to spot because he has a big kink in his tail. Once we'd seen the distinctive kink, we noticed that others in Nana's family had a slight kink to their tails too. It is a genetic trait that they inherited, along with her calm nature and good manners.

Whereas Nana's family members are generally polite and well-behaved, Frankie's descendants are more mischievous and the little ones can be downright hooligans until their elders take them in hand. Frankie's family also have the matriarch's beautiful long straight tusks.

Elephant tusks are very useful implements – they can be used to dig for water, lift logs and strip bark from trees. They are handy in a fight and also protect the trunk and face from damage. Just as we are right or left handed, elephants are right or left tusked. They favour one side more than the other, and the tusk on the dominant side is usually worn down from use.

Some elephants never develop tusks. Tusklessness is quite a disadvantage, particularly to males, when it comes to fighting. As a result, the tuskless males live and breed less successfully, and the gene for tusklessness has been naturally selected out in males. In a horrible evolutionary twist, in more recent decades, tusks have become something of a liability. Poachers prefer elephants with bigger tusks, because it means more of their valuable ivory to sell. As a result, the genes of the large, breeding-age males with big tusks are removed from the gene pool.

During the fifteen-year civil war in Mozambique, fighters on both sides slaughtered elephants for ivory to finance their war efforts, or for meat. An estimated 90 per cent of the elephants in the region that's now Gorongosa National Park were killed. Many of the survivors are tuskless – they never developed tusks. In that particular time and place, not having tusks put elephants at an advantage, so they stayed alive and bred. In a relatively short period of time, evolutionarily speaking, poaching has had a significant impact on the gene pool.

'Who's that big one with only one tusk?' asked the photographer, without taking her eye from the viewfinder.

'Gobisa. He's the oldest male,' said Andrew. 'Gobisa was a captive elephant, used for elephant-back safaris. If you saw him today, you would never know it. He has been success-fully rewilded. He was brought in to show the young bulls at Thula how to behave. Without a strong male role model, those young guys can get quite troublesome.'

'Remember when Mabula first met Gobisa? How upset he was?' Victor said with a laugh. 'He wanted to know, who was this big new bull at Thula Thula?'

'He was not happy,' I agreed.

Victor continued reminiscing: 'But Gobisa thought, "*OK guy, we can do this the easy way, or we can do this the hard way.*" And what do you think he chose?'

'Go on, tell me. What did Mabula choose?' she asked, lowering her camera and leaning forward to hear the outcome.

'He chose the hard way!' said Victor, with a burst of laughter.

'It was a rumble in the jungle,' I said, remembering the two great beasts going head-to-head, the sound of their trum-peting splitting the air. 'The ground shook!'

'How did it end?'

'Gobisa put his huge trunk on Mabula and pressed him all the way down to the ground,' said Victor, demonstrating with his arm as a trunk. 'Do you know, *gobisa* means "to bend", in Zulu? And that's what he did. He bent Mabula to his will. Gobisa was the boss. And that was the end of that. Until the next time.'

'There was a next time?' she asked eagerly.

'Yes.' Andrew picked up the story. 'A couple of months later Mabula came into musth. That's when a male elephant gets a big rise in testosterone and it can make them very aggressive. Mabula decided it was time for the return match. This time, he won. Gobisa was no longer the dominant bull of the Thula Thula herd.'

'Wow, it's like a soap opera.'

'It is!' I said. 'The ups and downs, the relationships and romances. Danger, intrigue, heartache and happy endings. Never a dull moment with our elephants.'

As the sun dipped and the evening star emerged from the darkening sky, I watched our happy herd and gave myself over to their joy. I felt a deep peace in knowing that they were where they were meant to be, living happy lives. They give me hope and comfort, they inspire me to be a better person and to try harder, they remind me of all that is good in the world. The elephants never disappoint. To be amongst these majestic creatures as they go calmly about their day is an extraordinary privilege and joy. I never tire of it.

I knew that I would fight for them to live this life, no matter what the future held.

When I think back to my first encounter with wildlife when I first moved to South Africa thirty-three years ago, I cringe with embarrassment. I was a city girl who had never met a wild animal that wasn't in a zoo behind bars. In fact, I had a great fear of dogs, and gave the little Parisian mutts a wide berth when I passed them on the street. My friends from those days can't believe it when they see me surrounded by all my dogs, big and small. The first dog that Lawrence and I adopted – Max, a male Staffordshire bull terrier – cured me of my fear and made me love them.

So, at that time I knew less than nothing about animals, but my friend Anne was visiting from France with her eight-year-old son Benjamin. He was a great animal lover, who delighted in watching the playful monkeys in the garden of our house, so I took them to a game reserve.

'Let's get a bunch of bananas,' said the boy. 'We can feed the monkeys.'

I thought, OK, why not?

With our bananas at the ready, I drove into the reserve. I gave the printed indemnity form and instructions a cursory glance – pages of very small writing about what to do and not to do, all in English – and tossed it into the back of the car. We had a happy day driving around the reserve, Benjamin on the lookout for monkeys and baboons to feed. We came upon a couple of rhinos and stopped to take photos. They looked very placid. Like cows, really, just bigger and greyer and in Africa.

'Get out of the car,' I said to Anne. 'I'll get a nice shot of the two of you with them in the background.'

She and her son did just that, happily posing in the grass, a few tons of rhinoceros grazing behind them, while I snapped away.

A car pulled up and an angry looking man leaned out to tell us to get back in the car.

How rude! Who was he to tell us what to do? Well, we told him to mind his own business, and he drove off in a huff. Afterwards, I realized that he was probably a guide driving tourists around and trying to save us from getting ourselves killed by the wildlife.

When my photos came back from developing a week or two later, I proudly showed them to Lawrence. When he saw the photograph of the two of them grinning in front of a couple of rhinos, he went pale, and then red!

'What were you thinking, letting them get out of the car?' he demanded. 'These are dangerous wild animals. You could all have been killed.'

I didn't mention the bananas.

I think about it with horror, now that I've been in the bush for twenty-two years. It was just dumb luck that we three ignorant foreigners – two blondes and a little boy – didn't encounter elephants or lions. We could have been killed, and the animals would no doubt have paid the price for our stupidity. I think people are a bit more informed about wild animals these days, but I'm still not a great believer in self-drive safaris in the big reserves.

We would have had a much safer and better bush experience with a guide driving and showing us around, just as they do at Thula Thula. Our rangers give a little speech before every game drive, reminding people that these are wild animals, and explaining what to expect and how to behave when encountering big game – don't make sudden gestures or loud noises, or try to pat them or stroke them if they come close. The rangers share their deep knowledge of the bush and the animals, which makes every drive a fascinating education.

Meeting our elephant family in the wild can be a deeply emotional experience. Popping into the Elephant Safari Lodge one evening, I saw a lady sitting alone on a sofa. She waved me over and introduced herself as Linda.

'I want to tell you about the most extraordinary, life-changing thing that happened to me on the game drive this afternoon,' she said. I could see that she was feeling very emotional, her hands shaking a little as she raised her glass of water to her lips.

'Please do tell,' I said, sitting down across from her. I love to hear our guests sharing their stories about their encounters here on the reserve.

'I have been terrified of elephants all my life,' she said. 'I've had a couple of scary experiences, mock charges. I almost didn't come when I was invited here to Thula Thula, but it was my cousin's special sixtieth birthday treat, so I came. But I was scared. I knew there would be game drives, and elephants . . .' She gave a little shudder.

The truth is, I understood her reaction. As much as I adore the elephants, ever since that day when Frankie nearly killed me and Lawrence, I'd been nervous when we met the herd up close on game drives. The trauma of that experience never left me.

'We went out this afternoon. Muzi was our ranger. He told me to sit right up front next to him and he promised there was nothing to worry about, that it would all be fine.'

'Muzi knows the elephants, and he knows people,' I said with a smile. 'You were in good hands.'

'I sat up front, still quaking with fear. We hadn't gone far when he spotted the herd up on the road ahead and drove towards them. He stopped the car. My heart was pounding, but Muzi said, "Just relax. Look how peaceful they are." It's true, they were munching on the nearby bushes and they did seem quite calm. In the side mirror, I saw one of them, a huge one – I mean extra huge, the biggest elephant of the lot – start walking towards us. This huge elephant walked alongside us and stopped next to me. I wanted to cry. Muzi said quietly to me, "It's OK," and to the elephant, "Hello Gobisa, how are you doing?" He came so close I could see the bristles, the wrinkles, the specks of dust. And then – oh, my heart! – he put his trunk out to me and touched my arm. I thought I might have a heart attack! But it was like a gentle kiss, as soft as a feather. I felt his breath on my skin . . .'

Her eyes were wide and filled with tears. I could see that she was reliving a very powerful emotional experience. She smiled. 'Suddenly I wasn't afraid any more. I was filled with peace and a kind of . . . healing. I looked into his eyes and it was as if we recognized each other. I know it sounds mad, but I believe he sensed my terror, something that had plagued me for years, and that he came to me to help me get past it, to show me what elephants can be like. The fear – it's gone.'

As she finished her story, she blushed slightly and said, 'You must think I'm crazy . . .'

It was a truly remarkable story. The elephants at Thula Thula trust us humans and like to come and see who is on the game drive that evening and what we're up to, so they do come close to the vehicles, but it's unusual for one of them to reach in and touch someone like that.

But crazy? Not at all.

Having lived two decades amongst elephants, I know that they are very intelligent, deeply emotional creatures with incredible intuition. It seems quite possible to me that Gobisa

– the big bull who had pushed Mabula to the ground with his great trunk – might recognize that this woman was in emotional distress and use that same huge trunk to tenderly 'kiss' her to ease her pain.

Elephants are remarkable, but at risk of sounding like a typical doting mother, I have to say that the Thula Thula elephant herd is unique. Many of our South African guests are bush lovers who have been to game reserves all over the country, and they all say that they never had such extraordinary encounters with elephants. Our elephants radiate contentment and trust, and they display it when they meet our guests on safari. They have a magical effect on humans, and I have heard many stories – like Linda's – attesting to powerful, transformative interactions between humans and elephants.

In the words of David Sheldrick, who was the founder warden of Tsavo, Kenya's largest national park, and who observed and studied elephants over many years: 'In order to interpret elephant behaviour, you must simply analyse it from a human point of view and that way, you will usually end up close to the truth, something the scientists have yet to learn.'

David's wife Daphne Sheldrick, in her book *An African Love Story: Love, Life and Elephants*, observed, similarly 'Elephants are, indeed, just like us, and in many ways, better.' She referred to them as 'the most emotionally human land mammals' and it is true they share many of our emotions. It is almost as if this common emotional range connects us to each other, humans and animals, in some mysterious way. It's as if we recognize some profound connection between our species.

Elephants do seem to display empathy. They recognize each other's pain and distress and respond to it. They even try to help. If you watch those wildlife YouTube channels – I do, even though I live in the bush! – you might have seen elephants helping a calf who has fallen down a riverbank, or who was attacked by a predator. Even if the calf is unrelated to them.

Young elephants like to jostle each other and fight amongst themselves in a playful way, testing their strength and establishing their individual identities, just like young humans. When the play fights get a tad too aggressive, adult elephants will put a stop to them, often pushing the youngsters away from each other. They even put them in 'time out', just as human parents do.

Marula, as the stern mother figure, is often the first to come over and separate the over-zealous combatants. Human mums would recognize the body language and her message: 'Right, you lot. That's enough of that. Now take it down a notch before someone gets hurt! You, leave your brother alone, get over here.'

What is interesting is that scientists have observed that the adults are not simply responding to signs of distress or pain, like a particular cry or sound. They intervene before the little ones get hurt or upset. They are thinking about the calf's emotional state and anticipating their future distress. Elephants think about each other's feelings.

In a matriarchal society, the mothers have a lot of sway

– and they're not afraid to use it. In recent months, Mandla, our biggest elephant, took a shine to little Andile. At just ten years old she was too young for him, but she appeared to be flattered by his attention, and the two of them seemed to have fun 'flirting' – walking together, reaching out a trunk to touch the other.

Mandla tried to take it to the next step, to cover this youngster, and mate with her. That wasn't what Andile had in mind at all! She tried to move away, but Mandla got a bit too insistent, and this was when his mother Nana and the aunties stepped in. With Nandi and ET at her side, Nana marched straight up to him and pushed him away from the younger elephant. The other females gathered around Andile to protect her.

For all his size, Mandla is a beautiful, soft-hearted soul, gentle like his mother. He obeyed his mum, and wandered back to the other young bulls, where I imagined him sharing the story of being rejected by the object of his desire.

If elephants recognize distress, and take action to prevent it, is it so far-fetched to imagine that Gobisa recognized a human emotion that day, and intervened to make our elephant-phobic guest feel better? I think not.

Elephants often seem to know when a person is scared of them, and make a beeline for that person. A lady from England, out on her first game drive, was so astounded by the size of Mandla in real life that she gave a whimper of terror and slid down into the footwell between the seats. I have to say that being a couple of metres from Mandla with him towering above you is quite an experience. He is one big boy. And she wasn't to know that he was such a softie.

Well, her attempt at disappearing had the opposite effect. It piqued Mandla's interest.

'Hmm, I wonder what she's doing down there,' he seemed to be thinking. 'Maybe I should investigate.'

He wandered closer to get a better look, further terrifying the poor woman crouched on the floor of the vehicle.

'Humans. What an odd bunch they are,' he thought, peering down at her, and giving her a sniff with his trunk. 'Ah well, not much to see here.'

And with that, he turned and went calmly about his business. Our guest returned to her seat to enjoy the rest of the drive, and returned to England with a tale to tell.

On chilly afternoons, we give guests rugs to keep them warm in the open vehicles on the game drives. On one particular drive with a rather anxious guest, the playful Mabona reached in to remove a lovely animal print blanket from the trembling guest's lap. She had loads of fun tossing it onto her head, then dropping it to the ground to kick it around a bit, then lifting it again and draping it onto her butt. Who knows how she came up with that circus routine, but she was definitely hamming it up for her audience.

You probably wouldn't imagine that an elephant has a sense of humour, but Mabona – named after our Elephant Safari Lodge manager, Mabona Mthimkhulu, one of the first people we hired when we bought Thula Thula twenty-two years ago – definitely does. She has a cheeky side, inherited from her mother ET (short for Enfant Terrible, which tells you all you need to know about her). She's playful, a born comedian, and she loves to show off.

When I want to give myself a good laugh, I go through my photo file for Mabona. Here she is doing a perfect Downward Dog that would win the approval of any yogi. There she is climbing up a muddy bank with one hind leg up but her big butt and her other leg hanging over the edge, as if waiting for a push from a helpful friend. In this other picture, she's sitting in the mud like a dog waiting for a treat. Really, she's hilarious.

I believe that one of the reasons that humans are so fascinated by elephants, and love them so much, is that they are so like us in many ways. They have their own personalities and quirks, their preferences and habits, their relationships and their deep social bonds. Charles Darwin wrote: 'There is no fundamental difference between man and animals in their ability to feel pleasure and pain, happiness and misery.' It is especially true of elephants, who share so many of our emotions.

3

A Risky Rhino Rescue

The clatter of helicopter blades split the peace of the Zululand bush. The chopper swooped down, coming in low, herding our two rhinos, Thabo and Ntombi, towards the open savannah grassland. I watched from the top of a nearby hill, nervous and on edge. Tranquillizing and dehorning our rhinos is a major undertaking and potentially risky – for the animal, and for the humans involved. But it has to be done to protect our rhinos against heavily armed and ruthless poachers who would kill them to get their horns.

The rhinos tried to outrun this great bird, but to no avail. As it drew close, the vet, Trever Viljoen, leaned out of the open side of the chopper with a rifle. I caught a glimpse of his slim frame, and his face, intently focused on his mark. He took aim and fired.

He fired a second time.

I didn't see the tranquillizer darts hit their target, but within minutes I could see the effects of the M99, a drug 1,000 times stronger than morphine. The two rhinos staggered and swayed. The vet's team and our rangers and volunteers ran towards them to control their fall and prevent them from injuring themselves. The rhino must come to rest on its chest and not on its side, so that it can breathe properly, and so that the vet can reach the horn to remove it. The assistant vet tied a cloth over each animal's eyes, and plugged their ears.

With shaking hands, an ashen-faced young volunteer placed a tarpaulin under Ntombi's huge head, and another did the same for Thabo. The horns are so valuable – they fetch $90,000 a kilo on the black market in the Far East – that even the tiniest shavings of horn must be gathered up, not a sliver left to attract poachers. Larry Erasmus, our security consultant, had assembled a team of ten armed men, trained anti-poaching dogs and several specially reinforced security vehicles to keep watch over the whole operation, to deter poachers and keep us and the rhinos safe.

Trever got straight to work with a chainsaw, first on Thabo, then on Ntombi, cutting away at the horn with intense focus. As ever, I was calmed by his presence in these stressful situations. He is always meticulously prepared and gives our animals excellent attention, whilst answering my questions with patience and humour.

Although it's painless for the rhino, a dehorning is a brutal, noisy thing to watch. Morgan Dias Simao, a filmmaker from France, was positioned right behind our vet, his cameras trained on the sleeping giants and the swarm of attendants. This was Morgan's first experience of dehorning. I could see the tears in his eyes behind his sunglasses. I knew how he felt – the first time I saw the vet start up a chainsaw and go to work on Thabo's horn, I cried as I witnessed what we had to do to keep our precious rhinos safe. I had now grown somewhat used to it, as we dehorn the Thula Thula rhinos every fourteen months. A rhino horn is simply keratin, the same substance that makes up our hair and our fingernails. Just like our fingernails, it continues to grow, and poachers would think nothing of killing a magnificent animal for just an inch or two of regrowth.

The dehorning itself is surprisingly quick. Within minutes, the horns were removed. Each one was immediately weighed and the rhino's name and the position of the horn is written on the base in permanent marker – 'Ntombi Front' and 'Ntombi Back'. A little hole is drilled into the horn, a microchip is inserted and it is sealed with wood glue. The vet signs it off, and the horn and the vet's paperwork are stored in a secure vault, not on our premises. The microchip is registered with the wildlife authorities, who keep a record of all the horns, and issue us with a certificate. The microchipping programme allows them to keep track of legal horns, to link each removed horn to a specific rhino and to trace stolen horns.

A DNA sample from the horn forms part of the state's rhino DNA database. When poachers are found in possession of a horn, matching its DNA to a specific rhino on a specific farm can help police to build a strong case and get a conviction.

Getting the horns off the premises under armed guard was like a military operation. The guards were fast and efficient, their eyes scanning the bush and the sky as they bundled the horns into the vehicles for transportation to a secure location.

We were very aware that dangerous people would stop at

nothing – including murder – to get their hands on this valuable cargo. A rhino horn is entirely useless to anyone other than a rhino, who uses it to dig, forage, break branches, and for defence. And yet many people, particularly in the East, believe that the powdered horn will bring health, strength and happiness, and will pay a fortune to get their hands on it. Around 10,000 rhinos have died in the past decade to meet this demand.

While the security team dealt with the removed horns, the freshly cut stumps were sterilized against infection and oiled to prevent cracking. Trever injected the reversal drugs to counter the anaesthetic, and everyone moved safely away from the stirring rhinos. It was a relief to see Ntombi get up quite quickly and start moving away from the dehorning site. Thabo looked cross-eyed and disorientated, as if he were waking up with a bad hangover after a big night out. He walked slowly off into the bush beyond the vehicles.

The strong anaesthetic leaves the rhinos weak and confused for a while, so the rangers always monitor them closely for forty-eight hours afterwards, in case they get into trouble. When ranger Victor and the volunteers went to check on them some time later, Thabo came right up to the car. There was no sign of Ntombi which was strange, because the two were always together. Thabo leaned groggily against the vehicle so that it couldn't move. Victor told me afterwards that it felt as if Thabo had come to tell them something, that he didn't want them to leave.

When they found Ntombi, they realized why Thabo had been trying to keep them there. She needed their help. Still woozy, she had staggered into a drainage ditch that channelled the water run-off from the plain. She followed the drainage line, the thick mud pulling at her feet. The ditch narrowed to a point where Ntombi could go no further. Nor could she reverse. She was stuck. The sun was low on the horizon and evening was fast approaching. Ntombi was tired from the anaesthetic and from struggling against the mud. She didn't have the strength to help herself.

The rangers gathered around anxiously.

'What are we going to do? There's no way forward and it is too narrow for her to turn around.'

'Well we can't pull her out.'

'We can't leave her here. If it rains and the ditch fills up, she could drown.'

'If we make the channel wider ahead of her, and create a kind of ramp, she can get out. We need to dig,' said Vusi. 'Where's the TLB?'

He was referring to the tractor-loader-backhoe, which is used for digging and clearing jobs.

Muzi answered, 'It's at the other side of the reserve, it was being used at the quarry. It will take half an hour to get here.'

'Tell them to come, but we can't wait. She might injure herself. We need to start digging, to clear the way for the TLB when it comes,' said Vusi. 'And get hold of Siya. Tell him what's happened and get everyone out to help. And call the APU. We need all the manpower we can get.'

The volunteers ran to the vehicles for spades and picks and whatever else they could find. Luckily, Ntombi didn't panic. She was very placid, standing quite still and breathing heavily.

She knew that she was in trouble, and that we were helping her.

'Where's Thabo?' Khaya asked Richard the rhino monitor, voicing the question that was on everyone's mind. This could turn into a very dangerous situation. If he saw his companion in distress, and a dozen humans gathered round, his instinct would be to protect her.

'He went the other way, past the vehicles. I don't see him. I'm watching for him though,' said Richard.

'Good. Keep a lookout. We're going to get to work.'

It was all hands on deck, using every tool they could find to clear the bush around the ditch and get digging. Nina, a young volunteer from Austria, and the daughter of my friend Heli Dungler, the founder of Four Paws, was with them. She was a tiny, cheerful young woman, a vegan who the rangers liked to tease – they called her Miss Chickpea – whilst enthusiastically dividing her steak between them at dinner time. She was always willing and ready to help out, and she was the one bravely holding a torch so the diggers could see what they were doing.

The Anti-Poaching Unit (APU) arrived with their guns in case of trouble. Morgan had hitched a lift back with them, and in the rush had only his phone and one camera. He looked at the activity, at all the people digging by hand, with such focus and intensity and asked, 'Why don't we get a helicopter and pull her out?'

Victor stopped digging long enough to give him a wry smile and say, 'A helicopter? From where? It would take two hours to get here and it would cost tens of thousands of rands. Anyway, it couldn't come into the bush in the middle of the night. It's too risky. No, we will have to do this ourselves.' And he went back to his task.

Morgan stationed himself behind the group to take pictures. The digging was hot, sweaty work and one of the APU guys handed Morgan his rifle, saying, 'Please hold this while I take my jacket off.'

Morgan took it gingerly. He had never touched a gun in his life. As he stood there with the rifle, pondering what a strange position he had found himself in, deep in the African bush, holding a gun while men dug a rhinoceros from the mud, he heard a noise. He turned to look. Thabo was ten metres from him and coming fast.

'Rhino charging!' Morgan shouted.

It was pandemonium!

Every book, every ranger will tell you not to run from a wild animal unless it's absolutely your only choice. This seemed like that moment.

Everyone ran. They didn't know where to go, but their instincts told them they had to move. Thabo was coming from the direction of the vehicles, so there was no hope of shelter there. They were in an open area of scrubby savannah grassland, and the only trees in sight were small, thorny acacia and prickly sickle bush. Nothing you could climb, or even hide behind. People scattered in all directions.

Victor threw down the saw he was using to clear a sickle bush. He started to run but was knocked over in the chaos. Scared volunteers and rangers ran right over him and Nina, who was still holding her torch.

Morgan ran in a zigzagging movement with the rifle in one hand and his iPhone in the other, with only its little light to show the way.

He told us afterwards, 'I was so scared! I turned and saw Thabo, perhaps two metres behind, running after me. Ahead of me there was light – it was Nina's torch – and I realized it was where we had been working. I ran towards it. I know that rhinos can't jump, so I jumped right over Ntombi in the ditch.'

On the far side, Victor and Nina were down on their knees in the mud. Morgan joined them.

Thabo stood just metres away, Ntombi in the ditch between them.

They had nowhere to go.

Nina was muttering quietly, 'Thabo. Stay there Thabo.'

No doubt the soulful Victor was sending his calming energy to the animal he'd known as a baby, willing him to keep his peace.

But Thabo was angry. Victor, who knows rhinos, could see the warning signs of an imminent charge. Rhinos have good hearing, and their ears swivel independently of each other to pick up sounds from all around. When a rhino is in danger, or he sees an enemy, his ears point straight ahead at the source of his unease. Thabo's ears were facing forward, pointing directly at Victor, Morgan and Nina. His tail was erect and curled up like a pig's, in anticipation of a charge. He made the harsh puffing noise that signals aggression.

And then, from behind the furious rhino, came the sound of a whistle and Muzi's voice, calling softly, '*Woza*, Thabo.'

Come, Thabo.

The words the rangers had used when he was a youngster, calling him for his supper or to romp with them in the garden.

'Thabo, *woza.*'

Thabo's ears twitched, then swivelled to better hear the familiar voice. He turned towards Muzi in his vehicle, and took a few curious steps in his direction. Muzi moved, still whistling and calling, taking him away from Nina, Morgan and Victor, who were as still as statues, but for their rapidly beating hearts. 'I put my fear in my pocket,' Victor explained calmly afterwards. 'It fits very well there.' Thabo followed the sound, and they breathed a sigh of relief as he moved away from them. They could see him in the bushes, where he had come to a stop just metres from them.

'What are we going to do now?' asked Morgan, who was shaking with terror.

'We carry on digging,' said Victor. 'We have to. Ntombi is getting cold. There is pressure on her diaphragm. If we don't get her out, she could die. Morgan, keep your torch on Thabo. If he comes, throw mud at him and shout at him to keep his distance.'

Now Morgan looked sceptical as well as terrified. Mud? Against a rhino? What kind of strategy was that? But he did as Victor said. Thabo circled the area, keeping a close eye on the proceedings. But he kept his distance.

Everyone got back to work with shaking hands and pounding hearts. Finally, they broke through, opening the drain enough for her to turn around and then get out.

'Go on, Ntombi, you can do it!'

She did something very unusual for a rhino: lifted up her front feet and scrambled over the edge.

'Yes, we did it! She's out!' Everyone screamed with joy, momentarily forgetting their exhaustion.

But it wasn't over.

Ntombi ran the ten metres or so to join Thabo. She stopped and turned. Both rhinos stared at the humans. The cheering had stopped. It was silent, but for the sounds of heavy breathing, and chirping and clicking of the night insects.

Victor said quietly, 'If they both come, we are in a bad situation. Jump back into the hole, she won't come back where she was trapped.'

The two rhinos held their gaze for another moment, and then turned and ran off into the bush.

As they did so, there came the familiar rumble of the TLB arriving. Too late.

No one who was there that day will ever forget it. The emotional strain of dehorning our beloved rhinos, and then the drama of Ntombi's rescue, and then Thabo's arrival.

'That was a close one. How are you? Are you okay?' I asked Victor, when the crew finally returned to the Lodge for tea, or something stronger.

'Life,' he said, philosophically'. 'One moment you're alive, in another moment you're not too sure.'

Years later, Morgan told me: 'To this day, I remember it as one of the most amazing, craziest experiences of my life. It really made me understand the challenges for conservation with limited means. We dug Ntombi out by hand. She would have died without the dedication of the rangers, who spent hours, on foot, at night, with only pickaxes, elbow grease, and, above all, passion, to save her life. Our lives were in danger. We were out there at night next to an animal that could have killed someone if he had wanted to. Those guys are unsung heroes, risking their lives on a daily basis to save and protect these special souls from extinction. The means are limited but the courage is immense! The world doesn't know that this is what it takes to save the rhinos.'

4

Our Two-Ton Teenage Troublemaker

In July 2019, I returned from the French launch of *An Elephant in My Kitchen* to a uniquely African problem – a smashed and mangled Land Cruiser, the victim of a troublesome rhino.

'It looks like it fell off the back of a truck,' said Christiaan, the conservation manager, shaking his head, hands on hips.

'Thabo did that?' I asked in amazement. It was a mess, its mirrors hanging off, the tyres burst. It hardly seemed possible that an animal, even a rhino, could have done such damage to such a solid vehicle.

'Yes. It broke down yesterday. It was late so we left it overnight,' said ranger Muzi. 'We went to fetch it today, but Thabo found it first.'

'Why would he do that?' I said, 'That car looks as if it was bounced around the bush like a beach ball!'

'Ah, Thabo. You know what he's like. He just likes to move things, to push them around. Once he's got it moving, it's like he's won and then he's happy,' said Muzi with a shrug and a grin. 'I dropped my cap once and he came over to sniff it and he just started pushing it around with his nose and flipping it. He was having fun.'

He sounded like an indulgent uncle talking about a mischievous teenager up to harmless pranks.

'That's not the same thing at all!' I said crossly. 'This is a car, not a cap! We can't have Thabo smashing cars.'

This was not Thabo's first act of vehicular destruction. When the mood took him, he could be quite the hooligan, and our vehicles had the dents and scrapes to prove it. Thabo was accustomed to our olive green game-drive vehicles, but he didn't like other cars coming into his domain. He liked to give them a push or a toss, just to shake them up a little and show them who's boss in these parts. A fully grown rhino weighs two and a half tons and can run at fifty-five kilometres an hour. With that kind of power, he doesn't have to try too hard to damage your paintwork.

Our APU's cars were so battered and dented by Thabo's antics that they designed a specially reinforced Thabo-proof vehicle, which they dubbed Mad Max. The bulked-up car certainly looked impressive sitting high off the road, its sides strengthened with heavy steel plates, a thick black bull bar at the front to protect against foliage and rambunctious rhinos. The APU guys went off to check the fences, monitor the rhinos and look for signs of poaching activity, feeling quite pleased with the new vehicle. Thabo took one look at it and clearly said to himself, 'That's all they've got? They think that thing is bigger and stronger than the great Thabo the Rhino?' And he gave it a couple of enthusiastic nudges with his enormous head, dishing out some bumps and bruises to the brand new, supposedly indestructible Mad Max.

'Ah, Thabo,' said the rangers, with a chuckle, and – I suspect – a dash of pride.

I had some sympathy for their indulgence. We have all known and loved Thabo since he came to live with us as a sweet little baby. All baby animals are dear, but baby rhinos are particularly lovable with their soft hornless faces and chunky little bodies. They are charmingly playful, bounding around like puppies on their short stocky legs.

Thabo had been found alone in the bush as a newborn, no sign of his family. He came to us at just a few months old. He grew up amongst humans, even sharing a room with his beloved caregiver, playmate and stand-in mum, Elaine, and snuggled up with her on her mattress at night. She would rub and stroke his little face to calm him when he was anxious. It was so sweet to see the look of blissful peace coming over him. An enclosed outdoor area, called a *boma*, was his backyard.

The game rangers played with him like a bunch of kids in the park with the family Labrador, tossing a rugby ball around and letting him run after it, giving him cuddles, and playing chase. Thabo's little ears swivelled like antennae when Siya, our head ranger, whistled for him. He'd run over to greet his friend.

A few months after Thabo's arrival, little Ntombi arrived. At just five months, the poor thing had seen her mother killed by poachers. Months of care helped her overcome the trauma and to trust humans and feel safe in the world. The two young rhinos became the best of friends. In my romantic French heart, I hoped that their friendship would turn to love. I was already excited for my rhino grandbabies!

Because we didn't have a proper rehab facility, they grew up as part of the family.

And grew, and grew.

Imagine an exuberant two hundred kilogram toddler, somewhat lacking in social graces, and unaware of his size and strength. That was Thabo after a few months in our care. His energetic escapades were becoming increasingly destructive. Chaos followed him and furniture shattered as he bounded boisterously around the place.

'It's time to introduce them to the bush,' said Siya, when the rhinos were about eighteen months old. 'They need to be exposed to life outside of the *boma* if they are going to learn to live as wild rhinos.'

I felt a flutter of anxiety. 'I worry about our little baby Thabo out there with all the dangers of the wilds. He was raised by humans. He doesn't know how to keep himself safe. I hope he's OK.'

'Our baby Thabo is a big boy now, just look at him,' said Elaine, smiling indulgently at the animal the size of a large chest freezer. 'We'll start with some nice long bushwalks to get them used to the sights and smells of the bush and take it from there.'

She and the Thula Thula rangers took the young couple on regular walks, showing them around the place – the watering holes, the sweet grasslands, the bush and the trails. The rhinos stayed out for longer and longer periods, exploring their expansive new world, and spent more time out on their own before returning to the familiar *boma* at night.

It was both a happy and a nerve-racking day when they

went off and didn't come back home to us at night. They were on their way to becoming wild rhinos – just as Lawrence and I had wanted for them.'

Even after a year or two of freedom, Thabo still loved the company of humans and had a particular soft spot for the people he grew up with. He sought us out, coming to the main house or the Safari Lodge to pay us a visit. He would sometimes put his huge head on the bonnet of the game drive vehicle, just like he used to when he was a little chap looking for a kiss on the nose. He let Richard the rhino guard pick ticks off him. He would come to my house and lie down on the other side of the fence, and Clément and I would have a little chat with him. I honestly believe that Thabo thought he was one of us – or that we were just rather strangely shaped rhinos. He wanted to be with humans.

He was too used to humans for his own good, and I worried that it would make him easy prey for poachers. Thabo also became more inclined towards destructive pranks – like challenging vehicles to a fight. We had to admit that Thabo wasn't a sweet little rhino calf any more. He was a territorial adult rhino – as he should be. He was also two and a half tons of pure bulk, built like a tank. And he simply wasn't safe for humans to be around. We needed to treat Thabo as the wild animal he was, not our little orphan playmate.

'We need to keep our distance from Thabo,' Siya told the rangers.

'Ah, but poor Thabo . . .'

'It's for his own good as well as our safety. No more

touching him, talking to him, playing. No eye contact. He needs to be a wild rhino.'

They looked quite downcast at this instruction, but they nodded in agreement. They knew this was how it had to be.

From that time on, we drove past him instead of stopping for a hello. We ignored his approaches. From one day to the next, he found himself excluded from the 'family'.

Thabo did not handle rejection well. He didn't understand why his playmates had turned their backs on him. Thabo was a particularly expressive rhino, and I could see he was hurt and felt sorry for himself. He hung his head and looked with sad eyes at the guests and game rangers who were doing their best to follow the new rules and treat him as a wild rhino, not our loveable baby Thabo. I just wanted to hug him, but I knew I couldn't.

5

The Thorny Problem of Thabo's Horn

'There's something wrong with Thabo,' Siya said, returning from a drive. 'There's a wound at the base of his horn.'

The guards had noted that he'd been particularly bad tempered of late, chasing cars and making a nuisance of himself. And then there was the Land Cruiser incident. If Thabo was in pain, that might explain why he was behaving so badly.

'We must call the vet and get him here immediately,' I said. 'We can't wait. Remember what happened to Numzane.'

Twelve years before, our big bull elephant had become suddenly uncharacteristically aggressive. I know you're not supposed to have favourite children – or favourite elephants – but Numzane was Lawrence's favourite. And the feeling was mutual. And yet, he rolled Lawrence's vehicle in a fit of fury. He became a destructive, dangerous menace to people and property. After six months of this, Lawrence made one of the hardest decisions he'd ever had to make: to have Numzane put down.

When the bull was euthanized, we discovered that he had a big, painful abscess at the base of his tusk. His episodes of 'bad temper' had been caused by the pain. Lawrence was devastated and filled with guilt that we had not seen his ailment or treated it.

I did not want to repeat this terrible history, so I called Trever and arranged for him to come and check on our Thabo as soon as possible. Trever takes his responsibilities for our rhinos very seriously, and he immediately got to work arranging the operation. The next day, he arrived in his car ahead of the helicopter, so that he could prepare all the necessary instruments and medicines and be ready to go straight to work. Everything was in order when the chopper arrived at the airstrip. It touched down to fetch Trever and his darting equipment, and took off in search of the rhinos, who had already been located by the rangers on the ground.

Trever darted Thabo and Ntombi. As well as treating Thabo's wound, he took blood samples from both rhinos and sent them for testing, to get an overall indication of their health. We also wanted to check whether Ntombi was pregnant (we were hopeful!).

Trever suspected that the injury might have been sustained when Thabo hit something hard (the Mad Max vehicle, perhaps? Or the Land Cruiser?) and split the horn. The wound was badly infected, so Trever administered an antibiotic injection to clear that up.

'Do you think the pain of the infection might have been responsible for Thabo's bad behaviour?' I asked Trever, remembering Numzane.

'It could be,' said Trever non-committally. 'We'll have to wait and see if his behaviour improves when the antibiotics take effect.'

'Let's hope this is the end of his antics,' I said, ever the optimist. 'And let's hope that Ntombi is pregnant!'

The end result was that Ntombi wasn't pregnant, and neither was this the end of Thabo's bad behaviour. But it was, at least, the end of his pain.

The stories of Thabo are legendary, and when we gather around the fire in the *boma*, we often end up sharing them.

'Remember that time when he didn't get out of the way for Shaka? He's got no bush manners, our boy Thabo . . . And that elephant just kept on walking straight at him . . .' said Victor one night.

Encounters between Thabo and Shaka, our feisty bull elephant, are always entertaining. To begin with, I was terrified that they would fight and try to kill one another. But Thabo only messed with that big elephant once. Amazingly, Victor captured the encounter on video.

They came across each other on the road for the first time face-to-face:

'Ahem. I'm Shaka, and I'm pretty sure I have right of way here.'

'You think so?'

'I do.'

'But I'm Thabo the Rhino.'

'Well, I'm an elephant, and as you can see, I'm fairly large, so . . .'

Now Thabo is no lightweight, but put him next to a huge bull elephant and he starts to look, well, I wouldn't say petit, but I wouldn't back him in a fight.

With Shaka looming above him, twice his size, and moving

determinedly forward, Thabo reconsidered his bold approach. Yes, Thabo thought better of going nose-to-trunk with Shaka. He turned round and trotted off on his way.

Since then, if Thabo crosses paths with the elephants, he keeps his head down, grazing peacefully. He treats them as he might treat a herd of antelopes passing by. Now, on other game reserves, elephants – especially young bulls – might kill rhinos if they meet in the wild. But at Thula Thula, they meet, greet, and move on.

We love Thabo, even when he behaves like a bit of a hooligan. We're like the parents of a rowdy two-and-a-half-ton teenager – much as we love him, we know that he's got to grow up and learn to behave himself. We needed to find a way to help Thabo adjust to being an adult rhino in the wild. But how? You can't Google, 'What to do with a troublesome rhinoceros.' Or even pick up the phone and ask some expert, or someone who has been through the same problem.

It was at times like this that I missed Lawrence's wisdom. He had so much insight and experience into the animals and the bush. I felt sure that he would know how to deal with Thabo in a way that was compassionate and kind and firm. He would say to me, in that calm way he had, 'Now Frankie, let's think about this . . .'

Nothing in my early life had prepared me for the difficult decisions I had to make every day in the bush. I grew up in Paris. I knew my way around the arrondissements and where to get the best pain au chocolat for breakfast. I had no formal training in conservation. Everything I know, I learnt on the job, from Lawrence or from people around me.

When we started Thula Thula, we built it from scratch. Money was a constant challenge, so we had to be really hands-on and do a lot of the work ourselves.

I took care of the hospitality side of things. To start with, that meant working with the builders to create the Elephant Safari Lodge, and then, a few years later, the Luxury Tented Camp. I furnished and decorated each room and luxury tent myself on a tiny budget, I improvised and I did it with love, style and loads of animal print fabrics. It just shows that when dropped into the deep end, you can do things that you never thought you could do.

Part of my role was to employ and train up people from the local communities in hospitality, food, reservation and marketing. Because I was in charge of staff, I took care of paying wages. In those days, most of the staff didn't have bank accounts, and besides, there was no such thing as paying electronically. I went into the bank in Empangeni and drew the wages in cash. The first month, as I was heading into town for the money, Lawrence called after me, telling me to take my little handgun, in case someone tried to rob me.

'I can't go into a bank with a gun!' I said. 'I'll be locked up.'

'Arrested?' he said, disbelievingly. 'No, don't you worry. It'll be fine, take the gun.'

He was so insistent that I did as he told me – after all, what do I know about South African banks? When I arrived, I was astonished to see all burly farmers marching into the bank with their big guns in holsters on their belts, in full view. My tiny firearm tucked into the bottom of my bag was a toy by comparison.

So in between choosing fabrics for the curtains, teaching the cooks to make a good bechamel sauce, and running around Zululand with a stack of cash and a gun, I had my hands full. I had no time to involve myself in the conservation side of Thula Thula.

After Lawrence's death, I was suddenly responsible for big decisions around conservation, the game reserve, the maintenance and the animals. Decisions with serious consequences.

I have learnt a lot in my years here, and I have common sense and good intuition. Even so, I never make these decisions on my own. I surround myself with people who really know the bush and the animals – from the rangers, to the vets, to the conservation experts. I consult the people I trust, who have knowledge and experience, and the same passion and vision as I do. Everyone has their say. I listen. I learn. When I've heard from everyone, and we've discussed all the angles, then I decide.

I called a meeting of all the rangers and Christiaan, our conservation manager, to get everyone's input on what to do about Thabo's bad behaviour.

'The infection at the base of his horn has cleared up. He seems in good health; I don't think it's that,' said Siya.

'Maybe he just wants attention,' said Christiaan. 'He can't understand why no one wants to play with him any more . . .' said Muzi. 'Poor Thabo.'

'The trouble is, he never learnt how to behave,' said Victor.

'He didn't have a dad to teach him what to do and give him discipline.'

'Maybe we should get him a big male rhino to keep him in line,' said Khaya.

'Wouldn't they fight?' I asked.

No one knew the answer for sure. We were in a unique situation with our temperamental problem child. We had to work it out ourselves, think it through, and find the best solution.

'Maybe he doesn't need a male, he needs a female,' I said.

'He's got Ntombi,' said Khaya.

'He's not hanging around with her as much. He wanders off on his own. Besides, they seem to have a brother–sister relationship, it doesn't look like they're going to mate and have a family,' said Siya.

'Do you think he might improve if he had a mate and a baby?' I asked. 'Settle down. Be a responsible family man.'

'Get Thabo a girlfriend? See if that sorts him out? I like it!' Christiaan said. There was a lot of nudging and laughter at the idea, but it seemed like a good one.

I decided we couldn't wait any longer. It was time to find a new girlfriend for Thabo.

6

A Girlfriend for Thabo

Finding your true love isn't easy. But finding a rhino sweet-heart for a tricky boy rhino with rather bad manners? Even more so.

There was no rhino dating site to help me in my quest, but if there was such a thing, I might have advertised: *'Lonely rhino male looking for a mature, experienced female, raised in the wild. For companionship, romance and more if there's affinity!'*

At eleven years old, Thabo was in his breeding prime. The rangers had observed Thabo scattering dung and spraying urine to mark territory. This scent-marking behaviour was a sign that he was ready to find a mate, so the time was right to look for a female for him. If our plan worked out, and we found the perfect mate, we would have a happier, calmer Thabo, and we would be able to add to the rhino population that had been so devastated by poaching. A double win.

In the absence of rhino Tinder, I called Simon Naylor, the general manager of Phinda Game Reserve, about three hours' drive from us. I knew that Phinda had a large population of rhinos, and hoped that they might be willing to part with one. I explained the situation.

'I am looking for a female a year or two older than him, someone with a bit of experience, to be a companion and a mate, to keep him in line and teach him about life in the wild. And hopefully make some babies!'

It did sound like rather a tall order, but Simon identified the perfect candidate – a mature female who had grown up in the wild. He knew she was able to breed and raise a calf as she had already produced a young female.

'The baby is twenty months old, so the mother is about ready to breed again,' he said. 'The baby will come with her, of course.'

She sounded perfect! It was just a question of sorting out the logistics – and the money. I sat down with Lynda to work it out. It's no trivial matter to transport two large animals, mum and baby, nearly two hundred kilometres. The list was long:

'The two rhinos will have to be darted, so there's the vet and the helicopter.'

'Right, and the vet will have to travel with them to check that they are stable.'

'Specially reinforced crates. And the truck to bring them here, of course.'

'And there's the security . . .'

Yes, security was a big concern and a major expense. Transporting two rhinos is like transporting a huge pile of cash. The poaching syndicates would think nothing of ambushing our trucks and killing people to get to two rhinos for their horns. We would need heavy firepower, a full crew of security guards and multiple vehicles.

'We're looking at about four hundred thousand rands,' said Lynda, dropping her pen onto the desk. 'We don't have that kind of money.'

'Well, we'll just have to find it,' I said.

Once more, we put our trust in our friends and in the people who loved animals. We launched a GivenGain campaign called Operation Girlfriend for Thabo. The message was simple – please help us buy a mate for our rhino, so that we can keep our boy happy, and make baby rhinos!

Mike, our international marketing manager based in Germany, had met Thabo on a trip to Thula Thula and fallen in love with our beautiful rhino. He got in touch and said he would do whatever he could to help raise money for Thabo's girlfriend. He helped set up the fundraising and promote it via social media.

People were touched by the story of the lonely male rhino looking for love. Once more, they supported us most generously. Within months we had the money we needed. Our two new rhinos would soon be coming to Thula Thula. What would we call them?

The naming of animals is always an important decision at Thula Thula and everyone gets involved coming up with ideas and suggestions. Every name has a story behind it.

Many of our animals are named for our human friends and family. Elephant Tom was named after our head chef at the Safari Lodge, who found the lost newborn. Suzanna and Jurgen are named after our dear Danish friends the Simonsens, who have been such great supporters of Thula Thula. The funny thing is, it's almost impossible to tell the sex of a baby elephant, so as luck would have it, Suzanna is a boy, and Jurgen a girl!

We were chatting about what to name our new rhinos when one of the volunteers, Maria, came up with the idea of Mona and Lisa, after the famous painting that we French call *La Joconde*. I loved the idea. This is one of the most valuable and beautiful paintings in the world, and very old. And our precious, prehistoric rhinos truly are a priceless work of art! It was decided. The mum would be called Mona, and the little calf, Lisa.

The day came. On 10 May 2019 Siya, Lynda, Vusi and I went to Phinda to fetch Mona and Lisa. Mike came with the film crew who were making a documentary. He had been so helpful in getting our fundraising campaign moving, it was only right that he be there to see the campaign come to fruition.

The mood in the car was one of great excitement, with a dash of nervousness. For Siya and Vusi and Mike, this was their first relocation. Even though we had employed a specialist relocation firm, Tracy & du Plessis Game Capture, we knew that a relocation was a huge undertaking. Darting the animals, transporting them, protecting them, unloading them. There was a lot that had to go right.

Even then, when we got them to Thula Thula, how would they like their new home? Would they settle in well? And how would Thabo react? I hoped with all my heart that he would love his new girlfriend.

We had to be at the gate at Phinda at 6 a.m., so we spent the night at a nice hotel in Hluhluwe, the closest town to the reserve. It was fun to be guests at a hotel for a change, and we were in high spirits, given our exciting task for the next day. Over a good dinner by the swimming pool, we discussed – of course – our new rhino girls. We were fixated on how Thabo might react, each of us playing out different scenarios, offering our own theories and thoughts on the matter.

'Well, we will soon know for sure,' I said.

We turned in early, ready for a 5 a.m. start. We were going to get our new rhinos!

We were at the Phinda airstrip bright and early. A helicopter and a game-viewing vehicle went in search of Mona and Lisa. The ground crew was in constant contact over the radios.

Word came from the helicopter: 'They are in an open area. We've got good visibility, no hazards on the ground. We're ready to dart them.'

We jumped into a Land Cruiser and went to meet them.

'Let's go!' said Siya excitedly. ' 'Let's meet our rhinos.'

The vet leaned out of the helicopter and shot two darts, hitting the mother, and then her calf. The rhinos were tranquillized enough to make them placid and woozy, but not enough to make them keel over. We didn't want three tons of fast asleep rhino! We needed them upright so we could guide the wobbly rhinoceroses into the heavily reinforced crates on their own four legs.

The rhinos darted, the ground team swung into action. Simon Naylor invited me to tie the cloth over the eyes of the adult female. I was excited to do the honours, and jumped out of the Land Cruiser eagerly. It was the first time I had done this task, and I was surprised to see that she was still awake and moving, although lying down. It was incredibly emotional and a little bit scary to get so close to this incredible creature to whom we were offering a new home. In my nervousness, I fumbled a little, and a young guy from the game capture team came to secure the cloth properly over her eyes.

While the rangers put mum and baby into the crate, I was looking around, glancing over my shoulder to keep an eye on the other rhinos that were around. When you've lived with Thabo the Great, you do learn to be a bit watchful when it comes to rhinos!

I was nervously looking left, right and behind, and had my eye on one male in particular.

'Why is this one looking at us so intensely?' I asked Simon.

'Probably her boyfriend,' he replied. 'The father of her calf.'

I didn't know if he was joking or not, but I suddenly felt bad.

'Oh no!' I said. 'Poor guy will be lonely.'

'Not to worry Françoise,' said Simon with a laugh. 'That's life in the bush. He will find another girlfriend.'

'How unromantic, Simon!' I said.

We were kidding around, but it made me think about how we humans make decisions on behalf of animals – to move them, or dehorn them, or put them on contraception – without their consent. Of course, we were doing this to help save an endangered species, and this lovely female rhino was coming to a good home at Thula Thula, where she and her baby would be monitored for their protection. But the bunny-hugger in me was a little discomforted. In a perfect world, without poachers or trophy hunters, rhinos would not need to be moved from place to place by humans, fenced in or monitored constantly to keep them safe. They would be able to roam free and wild and choose their own mates, and we wouldn't have to split up a family.

The process went smoothly, the capture team taking charge, with our rangers helping. The rhinos staggered into their separate crates, and the crates were winched onto the truck using chains. And we were ready to go.

The vehicles set off. A heavily protected bullet-proof security vehicle led the way, with marksmen at the ready, kitted out with every kind of firepower. Next came the truck with the precious cargo and more guards. Mike and the film crew followed in their car, followed by more security. Lynda, Vusi, Siya and I brought up the rear in our car. It looked like a presidential convoy!

We were approaching Empangeni, our local town just twenty kilometres from Thula Thula, when the vehicles in front of us slowed.

'What's going on?' asked Lynda nervously. 'Why are we stopping?'

My first thought was, *is it poachers? Or is something the matter with the rhinos?*

'Whatever it is, it's not good,' I said.

By now the whole convoy had pulled over to the side of the road, hazard lights flashing.

I checked my phone. There was a message from Mike: *Truck has broken down.*

Siya and Vusi jumped out. Their first concern was for the health of the rhinos.

'They've been in the crates for two hours,' said Siya. 'They must be wide awake now. I'm worried that they are stressed. And now they have to wait even longer.'

Vusi climbed onto the truck and pressed his ear and his hands against the crate. 'They are awake. I can hear them moving. At least we know they're OK.'

Meanwhile, Grant Tracy from Tracy & du Plessis Game Capture had begun calling everyone he knew in the whole of Zululand to assist us. And they came to the rescue. When Richards Bay Crane Hire heard there were rhinos involved, they dropped everything to help.

'The truck's on its way!' said Grant with relief.

Within twenty minutes, along came the biggest flatbed truck you ever saw. The game transport crates were loaded onto the new truck, and Mona and Lisa were back on the road to Thula Thula. It was a huge relief when the convoy reached the gate without incident. But there was another problem.

'How are they going to get that huge truck to the *boma*?' asked Vusi.

Good question. This truck was much bigger than the one we had originally hired. The narrow dirt roads on the reserve were not made for a massive flatbed. We didn't know if it would make it onto the reserve, let alone all the way to the *boma*.

To this day I can't believe that the driver, Lesly Ndawonde, managed to navigate the narrow tracks and then reverse that truck right up to the enclosure. If there's some sort of award for truck-driver heroes, he should get it. It was nothing short of miraculous.

The crates were deposited onto the ground and the doors opened. One of the rangers climbed onto the top of each crate to encourage the rhinos out.

Mona and Lisa walked out, and immediately moved to the centre of the *boma*. We all retreated to give them space, and took up positions far back from the fence so as not to stress them.

'Just look at Mona. Isn't she beautiful?' I said to Vusi. 'And sweet little Lisa. Thabo is a lucky guy.'

'I wonder where he is. I can't wait to see how he reacts,' said Mike, looking around hopefully. But there was no sign of him or Ntombi.

'He probably already knows they are here,' said Siya. 'He'll be able to smell them.'

The *boma* is a contained environment where the new arrivals can get used to the smells and sounds of their new home. It's close to the road from the Lodge to Tented Camp, so they get used to the sound of the game vehicles passing by, and the sound of people's voices. Because the *boma* is close to the Lodge, it's easy for us to check on them, admire them, and give them water and lucerne. The electric fence keeps the rhinos from getting out and keeps leopards and hyenas away – baby Lisa could be easy prey. They would stay there for two weeks to acclimatize, and then be set free.

Mona and Lisa were very still, almost subdued. I wondered

whether they were feeling the effects of the anaesthesia or of the long trip. A relocation is stressful. The rhinos had been separated for the first time since Lisa had been born and travelled hundreds of kilometres in a vehicle. And now they found themselves in this strange new environment with its unfamiliar smells and sounds.

A ranger gave the heads-up – we had visitors. Siya was right, Thabo and Ntombi must have smelled the new rhinos and came to investigate. I have to say, I felt a prickle of nervousness. Introducing new animals is very unpredictable, and none of us could anticipate how this would go down. Especially knowing Thabo . . .

At the arrival of the resident rhinos, Mona and Lisa moved closer to each other, as if for protection from some possible new danger. As for Thabo and Ntombi, well this was the first time they had met new members of their own species! The four rhinos kept their distance, watching each other and sniffing the air.

The rhinos were from two different worlds with a completely different upbringing. Mona and Lisa were totally wild, born in the bush, with no human contact, and probably no trust in humans. Our two little orphans, by contrast, were what we call 'habituated' to humans; in particular, to those who had taken care of them, fed them and looked after them. So, we all knew it was going to take some time for them to get acquainted . . . and it did indeed!

'Look at this beautiful expensive girlfriend we got you, Thabo,' I said. 'I hope you take good care of her.'

He and Ntombi turned and moved away into the bush and

didn't visit the newcomers again. The rangers couldn't wait to release them and see Thabo's reaction.

Thabo and Ntombi were nowhere to be seen when we opened the *boma* two weeks later. But the rangers kept a close eye, and before long, they saw Thabo approach the mum and daughter cautiously. Mona gave a short snort of warning when she saw him. Thabo seemed to be curious and excited, running about, but he gave her space. Although Thabo was bigger than Mona, he was younger and less experienced in dealing with other rhinos. Letting out the high-pitched sound that signals submissiveness, Thabo – all two and a half tons of him – turned tail and ran.

'He's scared of her!' said Siya.

Yes, the bold and fearless Thabo, who had once tried to attack an actual helicopter with its annoying clattering blades, who had got the better of the APU's souped-up Mad Max vehicle, was behaving like a timid teenager in front of a pretty girl at their first dance!

I felt sad for Thabo. Nobody had ever taught him how to behave in this situation. He didn't have a dad or an older male rhino to be a role model in matters of love and courtship. He grew up with humans. No matter how much we loved our Thabo, we weren't able to give him what a rhino family would give him. This was not going to be easy. He needed courage, and time to get to know Mona.

To keep our new rhinos safe from poachers, they needed to be dehorned. A month or two after their arrival at Thula, we arranged the procedure for Mona and Lisa. It was almost a year since the dramatic day when Thabo and Ntombi had been dehorned, and she had been rescued from the ditch, so we decided to have their horns trimmed too.

Trever darted all four of the rhinos from the air. Lisa, being the smallest, went down first. Mona was always very protective of her young calf, and when she saw Lisa fall down, she staggered over to where her baby lay motionless in the dust.

The tranquillizers were already taking effect on Mona, who was moving slowly and unsteadily as she approached Lisa. She raised one foreleg and stood there swaying, then fell to her knees. Two of Trever's assistants moved in with a cloth to cover her eyes, but in a last valiant effort, she got up again and staggered a couple of steps closer to her calf. Even in her drugged state, she felt the strong maternal instincts to love and protect her baby. As she lurched towards Lisa, the two men tried to push her away, but their strength was no match for her weight.

Mona was right next to her unconscious baby when the drugs kicked in fully and she started to go down herself. We watched in horror – the tranquillized two-ton rhino was about to drop onto her own calf.

'Stop her!' cried Christiaan. 'If she falls, she'll kill Lisa.'

The rangers ran to help the two men, all of them trying to keep Mona upright and steer her away from Lisa. It was like a terrible rugby scrum, with the rangers using their weight to prop her up as she threatened to crush her baby. They managed to hold her up, but couldn't move her.

When it became clear that they didn't have the strength to push her away from Lisa, everyone who was there that day ran to help. Even the film crew dropped their cameras and went to add their weight to the scrum. With an incredible team effort they managed to push Mona just far enough that she dropped safely to the grass, a metre or so from Lisa, who remained blissfully unaware of the drama and her brush with death.

The dehorning proceeded as planned from that point, and soon all four rhinos were safely horn free. The humans, on the other hand, were quite shaken by the experience. My heart was heavy when I came back to the Lodge. It is a necessary operation, but risky and stressful. I was relieved that it was over, and I knew that the rhinos would be safer without their horns. Even so, Thabo, Ntombi, Mona and Lisa would still have to be monitored by our rangers 24/7 and their location reported to me every thirty minutes. We knew of cases where even dehorned rhinos had been poached. Even the small stub of horn might be enough to attract poachers. This is the harsh reality of conservation in the face of human greed and cruelty. We can't afford to lose this battle to save an endangered species and ensure that future generations can see a rhino in the wild.

Now we had four healthy, dehorned rhinos, three of them of breeding age. How would their relationships develop? And would Thabo make us proud and do his bit for the survival of his species? Pressure, pressure, my poor Thabo!

Thabo might graze near the females for a little while, but would always move on quite quickly. This is normal behaviour for rhinos. While the females and the young stay together, male rhinos don't spend much time with them. The bulls are constantly patrolling their territory, marking its borders with dung and urine. When a female is ready to mate, the bull will pick up the scent from her urine and pay her a visit. But you might say that he doesn't stay for breakfast! Once they have mated, he will go on with his patrols.

While we waited to see whether Thabo would fulfil that role, we noticed an interesting change in him. Having grown up an orphan, Thabo didn't get the internal microorganisms that he should have. In the wild, he would have eaten his mother's dung and picked up the necessary good bacteria that way. But without the microorganisms, his dung was always runny. When Mona arrived, Thabo nibbled on her dung, and his own dung firmed up and started to look like regular, healthy rhino dung. Aren't animal instincts remarkable? We hoped those instincts would lead him to mate with Mona.

7

Rescue, Rehab and Release to the Wild

If I could have one wish, it would be for every wild animal to be safe and free in its natural habitat.

Unfortunately, that is not the case. There is always a need for animal care and rehabilitation. The sad thing is that humans are usually responsible for the suffering, whether it's training elephants for the circus, or keeping a bushbaby as a pet, or setting a snare that breaks a bush buck's little leg. Wildlife rehabilitation gives animals a second chance to live free.

In 2014, with the support of Four Paws, an international animal welfare organization, we created the Thula Thula Rhino Orphanage specifically geared to look after baby rhinos. The horrific and ongoing rise in rhino poaching has meant that baby rhinos are left orphaned, too young to survive in the wild, and often deeply traumatized by what they have seen and experienced.

We successfully looked after a number of baby rhinos as well as a baby elephant and a baby hippo. Our ultimate goal was always to release the animals into the wild, and for them to survive, thrive and reproduce.

In February 2017, five heavily armed poachers attacked the

orphanage. They killed two of our baby rhinos, only eighteen months old, to take their tiny horns.

I can't describe the pain, grief and rage I felt at their loss. The animals we have raised by hand, loved and sworn to protect, were slaughtered in their place of shelter. My innate belief in the good of human nature had been shaken by violence and betrayal. The orphanage we had started with the very best intentions, to help animals in need, had been tainted by tragedy. And then the organization in charge of the management of the orphanage took the decision to give up on it.

The orphanage closed.

In the weeks that followed, as I came back from the darkest place, I realized that the only way to salvage something from this horror was to create something positive in its wake.

'If people like us quit at every setback, there will be no more rhinos, no elephants,' I told the staff. 'We will rebuild what has been lost, and keep the dream alive. We will create a new animal care centre.'

The Thula Thula Wildlife Rehabilitation Centre opened in May 2017, three months after the tragedy.

The rehab centre was created to care for wild animals that have been injured, orphaned or taken from their natural habitat, with the aim of releasing them in the wild. This time it was managed by the Thula Thula team. We put in an application for a permit for a general rehabilitation centre – our previous permit was for rhinos only.

The centre was built on community land and had the support and blessing of the *amakhosi*. We are very fortunate that our local leaders are so deeply committed to conservation

– they take their responsibility to protect the land and its animals as a legacy for generations to come seriously. Four Paws also got involved again as our other partner.

The ultimate aim of everything we do is to send a healthy animal back into its natural habitat, prepared to live as a wild animal. That means that they shouldn't be held, handled, petted or cuddled by humans – no matter how tempting it is. They shouldn't get too comfortable with humans, or too reliant on us.

The work is often physically demanding and emotionally stressful – our rehabilitation carers see animals sick or in pain, and sometimes we lose them. But there are heartwarming success stories, like our first guest, Lucy, a common duiker. She was a tiny faun-coloured antelope – when fully grown she was no more than knee high – with a soft, pretty face and a shy demeanour. The local police found her captive, most likely to be sold as a pet or used as meat, and brought her to us. She was comfortably installed in our covered neonatal *boma*, where the new rehab centre team took good care of her, and bottle-fed her until she was able to graze.

A few months later, we released her on the lawn at the Safari Lodge. The name duiker is related to the Afrikaans word that means to dive, and Lucy lived up to her name. She came out of her crate and ran straight into the bush, ducking and diving into the protective cover of the undergrowth. We were delighted to see her natural instincts taking over – Lucy was ready for her new free life. We still spot her from time to time, and it makes my heart sing to see her so well and so wild.

Once word got around that the rehabilitation centre had

opened, local people started bringing us wildlife that they had found, or that had run into trouble. People often arrive with tortoises that have been seen wandering around the streets of Empangeni, where they are at risk of being run over by cars or attacked by dogs. Tortoises don't usually require much in the way of rewilding. We set them free at Thula Thula, and they make their slow and deliberate way into the bush, where – we assume – they live happily ever after.

We took in a bush baby who had been attacked by two Ridgebacks, with his back leg badly damaged. He was quite an old chap, who wouldn't be able to rewild entirely, but he lived out his days safely in a large *boma* at the centre.

Three newborn genets were found in a box at Richards Bay harbour. These small wild felines have adapted well to the urban and suburban life, feeding on whatever they can find, snacking on cockroaches and helping themselves to the pet food they discover conveniently lying around in bowls outdoors.

The mother was nowhere to be seen, so the little genets were brought to us. They were so tiny that they had to be fed with a syringe. The veterinary nurses fed them round the clock for twelve weeks, with minimal human contact. As they grew, these pretty long-tailed spotted cats spent more time outdoors in the *boma*. After a while we started to leave the gate open so the genets could wander in and out at will. They started to go further afield, and stay away for longer, until they didn't come back at all. Now they are wild genets, just as they should be.

Not all rehabilitations are successful, though. In fact, they can be heartbreaking when animals can't be saved, like a monitor lizard brought to us with a broken back, who had to be euthanized. Even quite healthy animals can die of shock after a traumatic incident or injury.

The trees around the volunteer academy are home to a colony of weaver birds. Weavers are busy little fellows, with cheerful bright yellow plumage. As you can probably guess from their name, the male weaver makes a nest from grass and reeds. These beautifully constructed woven nests hang down from the trees like lanterns or baubles. The female lines the nest with feathers or soft grass in preparation for her eggs and, hopefully, the chicks.

The diederik cuckoo takes advantage of these skilful crafters, and lays its egg in the soft and safe nest made by the weaver, essentially stealing it.

One afternoon, in the summer of 2018, the hatching season, a massive electric storm came over Thula Thula. It felt as if it was the end of the world. Lighting illuminated the sky, and rain lashed the ground. The wind howled and shook the trees. When the storm moved on and the rain stopped, we looked out on utter devastation. Leaves and branches and weavers' nests littered the ground.

'Oh no,' said Jack in horror. 'Look at all the baby birds.'

Little carcasses covered the ground, their fluffy feathers damp on their tiny bodies.

'That one's moving!' said one of the volunteers, pointing. 'It's alive.'

Amongst the forty dead baby weavers and cuckoos were fifteen tiny birds that were still breathing.

'We need to get them warm,' said Jack, scooping them up gently and depositing them in a cardboard box. 'Birds are very fragile and difficult to save, especially the little ones. I know they have a poor survival rate, but we have to try.'

Jack and the volunteers worked themselves to the bone feeding and caring for the little birds, but as the night wore on and the day broke, they died one by one. Only one – a cuckoo – made it through the next few days. The volunteers named him Chuck, and he thrived, eventually taking up residence in the trees inside the camp.

Chuck flew away one day – just as we would wish him to. Whenever we hear the characteristic *deed-deed-deed-deed-er-ick* call of the diederik cuckoo, we wonder if it's Chuck, the sole survivor of that devastating hurricane.

8

Saving Little Lives

In October 2018, Lynda took a phone call from a Durban number. 'We have a meerkat,' said the caller, who introduced herself as Jane. 'I hear you have a rehab centre. Can you take her?'

'A meerkat in Durban? How did you get her? There are no wild meerkats here in KwaZulu.'

'She was a pet, I'm afraid. She needs a new home.'

If there's one thing that makes us all quite crazy, it's wild animals being kept as pets. If you want a pet, get a dog or a cat. But not a meerkat or a monkey or even – believe me, it happens! – a tiger. Wild animals belong in the wild.

Because meerkats look so cute and they are not very big, people are sometimes tempted to try to keep one as a house pet, but meerkats are wild animals. It's not fair to take an animal out of their natural habitat and keep it in a home with humans. It's also dangerous. Meerkats can be aggressive and give you a nasty bite.

If you've seen meerkats in the wild, or on the TV show *Meerkat Manor*, you'll know that they are very social creatures. A group of meerkats is called a mob or gang, and hanging out on a sandy mound, they do look rather like a group of naughty teenagers looking for trouble.

Although they are on opposite ends of the mammalian size spectrum, meerkats and elephants have quite a lot in common in the way they live together and help each other for the good of the group. Like elephants, meerkats are matriarchal – there's one boss lady – and they cooperate with each other for the good of the group, taking on different roles.

When the meerkat pups are born, adults pitch in to babysit them so mum can forage for food. There will always be one meerkat acting as a sentinel, sitting up tall on its hind legs, its sharp eyes scanning the land and sky, on the lookout for eagles, snakes, jackals and other predators. While the sentinel keeps watch, the rest of the group looks for the insects that make up most of their diet.

If a predator or danger is spotted, the lookout sounds the alarm. Meerkats have an elaborate communication system of about twenty-five different calls. They all recognize the 'TROUBLE IS COMING!' call, and when they hear it they will either group together to face down the enemy, or dash to their burrows to take cover.

Imagine a social, communal animal like a meerkat, who should be in a group of twenty or thirty meerkats, living in a home with only humans for company. What a lonely life it must be.

Lynda, Christiaan and I drove to Durban to fetch the meerkat, whose name was Daisy. I had never seen a real life meerkat before and what an adorable animal she was, with her pointy little face and black eye patches. She was so alert and clever, her head swivelling side to side, looking about, taking it all in.

'Come on Daisy, let's take you to your new home,' said Lynda, when we got her to Thula Thula. We prepared a big enclosure for her inside one of the large *bomas*, to give her a sense of being out in the wild. Our goal was to release Daisy, but you can't just take a 'tame' meerkat and put her in the bush and expect her to survive. Daisy had never had the opportunity to learn how to be a wild meerkat, with all the skills that entailed. She would have to learn those skills before she could go into the bush.

The big enclosure was a good first step. The rehab centre staff were strict about keeping human contact to the minimum, helping her to unlearn her comfort with our species, and to start to do more for herself. We gave her what she would be eating in the wild – live insects – so she became used to foraging and catching them herself, as well as fruit, eggs and plants.

Daisy did become more wild and self-reliant in the rehabilitation centre, but we aren't rewilding specialists, and besides, we couldn't release her here – there are no other meerkats at Thula Thula, or even in KwaZulu-Natal. She would have to go somewhere where there were others of her kind.

Lynda likes nothing more than a challenge, and when there's an animal's well-being at stake, she goes into serious problem-solving mode.

'Good news!' she said, after Daisy had been with us for a few weeks. 'Solidearth Meerkat Rehab and Rescue will take Daisy and rewild her.'

'That's wonderful, where are they?'

'Kalahari.'

'That's the other side of the country!' I said.

'Yes. Up by the Botswana border. But that's where she should have been all along. Once she's there she's home.'

We both shook our heads at the insanity of taking a wild animal from the northernmost desert area of South Africa to the tropical coastal city of Durban as a pet.

Daisy was relocated to the Kalahari, back to her natural habitat. Elma and her team at Solidearth gradually rewilded her, exposing her to wild meerkats, so she could learn from them. After a month, she was finally able to join a group of meerkats out in the wild.

Not long after that, Lynda came walking across the lawn from her office to mine. As always, she was trailed by her big Labrador Miley, and the princesses of the pack, her two white poodles, Shiny and Alex, who had moved with her from Pretoria. I could see from Lynda's smiling face and her eager bustle that she had good news for me.

'Françoise! I've just heard from Elma,' she said. 'Daisy has had four pups!'

Lynda still keeps in touch with Solidearth, but she has no further news on Daisy and her pups. Our little meerkat rehab is now so well-integrated into the group that Elma can't tell her apart from the other meerkats. This is a very successful story of wildlife rehabilitation, a good outcome to a situation that should never have been created by humans.

*

The saddest situation, for me, is to come across an animal that has been caught in a poacher's snare. It is such a horrible, painful way for an animal to be injured or die. An ancient buffalo was very fortunate to have been spotted in just such a state, but alive.

We have about eighty buffalo on Thula Thula. Buffalo are one of the big five, and while they might not have the lion's fearsome fangs, or the elephant's impressive heft, they are considered amongst the more dangerous animals. They're prone to charge suddenly, and once they do, they don't stop. The old buffalo bulls, in particular, are notoriously bad-tempered. While the females live together in breeding herds with one or two males, and the rest of the males live together in bachelor herds, the elderly chaps are thrown out. They become grumpy old men, wallowing in mud, and ready to take offence at the least provocation.

Any buffalo that has lived on a hunting reserve should be treated with extreme caution – they've learnt to distrust humans, and are very unpredictable. Our buffalo at Thula Thula have never been hunted, so they are quite relaxed. In fact, over the years they have become more and more so, and now they come very close to the Lodge, where they feel safe. They've been making lots of babies – always a sign of a healthy, happy, well-fed herd, I believe. Mabona was surprised to see one actually inside the Lodge grounds, grazing along-side the nyala. But we did have a scary near miss when the original group of sixteen buffalo arrived in 2010.

The new arrivals came in a big truck. We watched from our vehicles from a distance, so the animals had space to get out and move into the bush. The ramp of the truck was lowered and out they came, thundering past at tremendous speed. We watched until they had all disappeared into the bush, and the vet approached the truck. As he reached the vehicle and peered in, a huge sleepy buffalo emerged. We hadn't counted them as they'd hurtled out, and had no idea that he'd been left behind in the truck.

The two of them eyed each other for a second, while we looked on in shock.

With a yelp, the vet took off at a run. The buffalo leapt out and went after him. Round and round they went, circling the truck. The vet was a little chubby and out of shape, puce in the face from exertion, but he put on a superhuman effort. The buffalo was at his heels, furious. Round they went. It was like a scene from a cartoon, and would have been hilarious if it wasn't so serious. A buffalo's horns are fused at the base to a hard bony 'boss' across the top of the head, and they curve upwards and outwards into sharp points. If they'd made contact with the vet, he would have been in significant trouble. The vet managed to climb onto one of the bakkies, hauled to safety by a couple of rangers. The furious buffalo ran to join the rest of the herd.

Who knows, the ensnared fellow we rescued might even have been the chap who nearly trampled the vet all those years ago. He was one of the oldest and largest bulls, so old that he had rheumy red eyes. The rangers called him Lucifer, and Siya once said of him, with a shudder, 'Those eyes! When he looks at you, you think he can end your life with just a mere thought.' We called in the vet and a helicopter, and with the land and air teams, lucky Lucifer was rescued, treated and set free to continue to kill mortals with his red-eyed stare!

Every time an animal is treated and released to live the wild life it is meant to live, my heart fills with joy, but sometimes there are bumps in the road. In February 2019, we asked for a permit to release a caracal and two genets, which had been rehabilitated in another facility, at Thula Thula. We were surprised to be turned down by the wildlife authorities. We were also still waiting for the registration permit for the rehab centre. Applications had been made, our facility had been inspected, but no permit was forthcoming. It felt as if we were being blocked at every turn, for reasons we never understood.

In March, we were asked not to accept any more rescue animals at the rehab centre until our permit to operate had been issued. The centre was forced to close. I fully understand the importance of rules and regulations, but it seemed tragic that our magnificent facility with its clinic and neonatal unit, its release *bomas*, its kitchen and storerooms, stood empty. We were ready and waiting to take care of animals and prepare them for life in the wild, but were unable to accept them. As there was no other rehab facility in our area, any wounded animal would have to be driven three hundred kilometres to the nearest centre, at significant expense, and with great stress to the animal.

It turned out that our first application to reopen the rehab centre wasn't done correctly, so in April 2019, we presented a new application, prepared by our lawyer this time. What haunted me was the knowledge that animals would likely die or be euthanized, when they might have been rehabilitated in our facility, and living healthy and free lives in the wild.

9

Christmas Bells and Smells

They say that there's the family that you are born with, and the family you create as you go through life. I have been blessed with my Thula Thula family – the animals and the staff and some treasured friends amongst the guests. We support each other, help and cooperate, and act for the good of the family as a whole.

Jo Malone and her family – husband Gary and son Josh – are honorary members of our Thula Thula family. Jo contacted me in 2019, having read my book. I knew her name, of course. Jo had developed a range of beautiful fragrances under the Jo Malone brand, and it had become an international sensation. She said she had never been to Africa, never been to a game reserve, but she wanted to come and visit Thula Thula, and she wanted to help the cause of conservation in Africa.

A couple of months after we spoke, Jo and her beautiful, down-to-earth family arrived and fell in love with our little piece of paradise. Something about Thula Thula touched them deeply. It was a million miles from the life they knew in London, but somehow it felt like home. We bonded instantly. She has a positive, can-do attitude that I just love.

While Jo was here, we talked at length about conservation and what could be done to protect animals and habitat. They witnessed the dramatic dehorning of Mona, Lisa, Thabo and Ntombi first-hand, so they understand what it takes to save rhinos from extinction. Having spent two weeks with us, and witnessed the work we do to protect and save African wildlife, Jo and her beautiful, caring family developed a passionate commitment to the cause. They do whatever they can to help bring international awareness to what was being done in South Africa to save rhinos.

In December 2019, I was in my office with Lynda, talking about our Christmas preparations, when I heard a commotion in the outer office. The three office ladies, Aphi, Portia and Swazi, came in, each struggling with a huge box.

'Christmas presents!' said Portia, grinning with delight. 'And there's more!'

They dropped the boxes on my floor and came back with three more.

'They come from the UK,' said Aphi, turning over the box to read the label. 'What could they be?'

Jo's assistant had asked me for our exact address prior to this, so I sort of guessed where the boxes had come from. But six huge boxes? Wow!

'Can we open them?' asked Swazi.

'Of course!' I said. There was no way we were going to wait. We were like kids on Christmas morning.

Each big box was stacked with smaller boxes, pure white, tied with a red silk ribbon. The little boxes themselves were simply magnificent – we all agreed that just the box could have been the present! Inside them were a selection of scented candles, the famous Jo Loves candles, and all sorts of beautifully scented bath and body creams and shower or bath gels. Jo could probably hear the delighted squealing from London, but just in case, we made a video call so that she could see the total ecstasy that her beautiful gifts had inspired. My office was in a hell of a mess, knee-deep in paper and boxes, but what joy we had that day!

There were exactly sixty products, one for each of our employees. I organized a Christmas party and let each of the staff choose a gift. No one at Thula had ever seen such luxurious products, and we had a wonderful time sniffing and testing, choosing and swapping. It made me smile to see our rough tough rangers discussing the exotic creams, lotions and candles all the way from London.

'Green orange and coriander,' said Muzi, his large hand wrapped around the delicate bottle. 'I like the sound of that.'

'Green orange? What's that? Oranges are orange,' said Khaya with a frown.

'Ah, what do you know about bath gels?' Muzi replied defensively.

Siya had picked up a tub of body cream and read the label. 'White rose and lemon leaves.'

'I'll have that,' said Victor.

'Oh no you won't. Choose your own,' said Siya, hanging onto his choice.

Jo gave us all so much joy that Christmas. Her generosity was just overwhelming.

By then, I had read her book *Jo Malone: My Story*, so I

knew about her life – the family struggles, her dyslexia, her terrible health challenges. She overcame so much to reach the great heights of success and fame that she did. Her courage, determination and resilience were an inspiration to me, as to so many others. But most of all, I appreciated her thoughtfulness and kindness to her Thula Thula family, who – thanks to her – were now celebrating Christmas in the deepest bush of Zululand with the smell of Mango Thai Lime lingering improbably in the air.

Christmas was happy and fragrant, but 2020 got off to a bad start. On 6 January, I got an early morning call from Vusi, our farm manager. A 7 a.m. call is always a worry, and his voice was grave.

'Françoise, the hyenas have attacked Lisa. We have been chasing them away, but Lisa is not looking good. She's lying down in the grass.'

'Is she injured?'

'We haven't been able to get close enough to see. Mona is standing next to her, guarding her from the hyenas. We can't approach.'

I put down the phone and immediately called Trever.

When he arrived a couple of hours later, we went straight to check on Lisa. The little rhino was still lying down in the grass, making sad little squeaky noises. A baby rhino in distress makes a terrible, almost human cry that goes straight to your heart. She was clearly suffering, but from what we didn't know. Mona stood helplessly by. Her maternal instincts led her to protect her baby but there was nothing she could do to ease her pain.

'I can't examine Lisa with Mona there. We will have to dart Mona.'

'If that's what we have to do, let's do it,' I said.

Luckily, we were in an open area of grassland, so the darting was quite simple, but Mona resisted the anaesthetic. It was as if she didn't want to go to sleep and leave her baby Lisa unprotected. Instead of going quietly down, she stumbled around. Eventually she succumbed to the drugs, and Vusi and the vet's assistant helped her to fall without hurting herself.

The vet examined Lisa. There were no obvious injuries. Trever took blood samples for testing. He gave her antibiotics and boosters, which is quite standard procedure when we're unsure what is wrong with an animal.

'I can't see any blood or wounds, so Lisa wasn't attacked by the hyenas,' he said. 'It seems that they arrived after she had already taken ill. I don't know what's wrong with her, but I don't want to leave her here in this state. She's too vulnerable to predators.'

'Let's take her to the rehab centre,' I suggested. 'She will be safe there in the enclosed *boma* and our rangers will be there to monitor her.'

It was a fine solution with one small problem – how to get her there? The rehab centre was close by – just a kilometre or so up the hill – but little Lisa already weighed about a ton and a half. I called Grant Tracy from the game capture company. When he heard about our rhino in danger, he said he'd get going immediately.

Grant arrived within a couple of hours. His team put Lisa

on the truck. The poor baby was still crying. It was heart-breaking and left us all feeling completely helpless. Once Lisa was loaded onto the truck and on her way to the rehab centre, the vet injected Mona with the reversal drug to wake her up. I felt so bad leaving Mona, such a devoted mother, without her daughter. I knew how stressed and sad she would be. But we had no choice, Lisa needed urgent attention and fulltime care. Once she was better, the two would be reunited.

Lisa was still unconscious when she was delivered to the *boma*. The APU rangers would watch over her.

'Let her sleep off the anaesthetic,' Trever said. 'She should wake up soon.'

'Please let me know as soon as she does,' I said to the rangers. 'And I want a report back every half hour.'

Lisa never did wake up.

I got the terrible call from Vusi at 2 p.m. 'Françoise, I am sorry. Lisa has died.'

I cannot find the words to describe how devastated I was, how devastated we all were. Our precious baby Lisa, gone, at just two years old, six months after she had arrived at Thula Thula. The blood tests came back. The cause of Lisa's death was a virus.

'It's so unfair. So inexplicable,' I said to Vusi, in tears. 'We go to extraordinary measures to protect our rhinos from savage poachers, and a bloody virus takes her away in just twenty-four hours.'

'It's nature. We can never understand . . .' he said, shaking his head.

I feel incredibly blessed to live the life I do, in the African

Frankie and Mabula.

The herd of twenty-eight Thula Thula elephants.

One of my favourite images of Frankie in all her glory.

Frankie in my garden!

Shaka and Thabo getting very close.

White-backed vultures.

Welcoming the arrival of Mona and Lisa with Siya, Lynda and Vusi.

Ilanga the elephant meeting Mona and Sissi. A really special moment, spectacularly captured by ranger Victor during the rhino monitoring.

(right) The buffalo of Thula Thula with one of our game-drive vehicles.

(below) The elephant herd interacting playfully with some of our guests.

Baby Jo the hippo.

Baby June the giraffe.

Mona and her calf Sissi.

Savannah – the first cheetah seen in the area since 1941.

Frankie and Gobisa – a special love story!

Mabona playing with a blanket.

bush with all its beauty and excitement, warmth and peace. But I don't think I will ever be able to quite come to terms with the brutal realities of nature's laws and the ways of the wild. They seem cruel, although I know that they are not. One animal killing another to eat isn't cruel, a virus isn't cruel – cruelty is killing an animal for its horn, or for sport, as humans do. But still, I feel each loss, each death, deeply and painfully. The loss of any of our beloved animals is sad, but to lose a rhino is particularly so, because they are so endangered, and their survival so precarious.

If we humans were upset, what about Mona? She was alone now. She had come round from her sleep not knowing where her daughter was or what had happened to her. Now, she was wandering like a lost soul in the northern part of the game reserve where she and Lisa had made their territory. She was, in fact, quite close to the rehab centre where her daughter died. Would she somehow sense her daughter had passed? I wondered. Or was she anxiously looking for her?

But the most amazing thing happened.

Ntombi had been very near to where Lisa had fallen sick and had observed the whole event from a distance – the darting, the relocation of Lisa, Mona waking up alone. Ntombi had tended to hang out with Thabo, but now, she dumped him and went to join Mona. She stuck by her to keep her company and comfort her in her time of loss. We were overjoyed to see the two females together. Ntombi took Lisa on a tour of the whole of Thula Thula, showing her the vast land which was now her home. Mona had never been in this part of the reserve before, so it was a discovery trip for her. They even came past my house, walking slowly by, shoulder to shoulder.

'I wonder what rhinos chat about on a girls' trip,' I joked with Lynda, as we enjoyed the sight of the two large ladies strolling down the road.

'Oh, I imagine Ntombi pointing out areas of interest. Like your house – that's where the boss lady lives.'

'I hope Ntombi has nice things to say about me!' I laughed.

Of course, we don't know how animals communicate, or what they say, but I imagined Ntombi somehow letting Mona know that she was not alone, and taking her on a little walk-about to lift her spirits. It soothed my heart to see the compassion and female solidarity that Ntombi demonstrated to Mona in her time of grief. It was truly beautiful to witness.

But what happened next was quite extraordinary.

Two months after Lisa died, our rhino WhatsApp group received a crack-of-dawn message from Muzi, who was monitoring the rhinos. It was a photograph of Mona with a tiny baby next to her with the comment: *We have a new baby rhino!*

We were stunned. We had no idea that Mona was even pregnant. A rhino's gestation period is sixteen months, and it's almost impossible to notice a pregnancy in a large creature with a big tummy. With rhinos, as with elephants, the first indication we have of pregnancy is often the arrival of the baby! To say we were overjoyed would be an understatement. It was as if the universe had given us this amazing gift after the terribly sad loss of Lisa, which we still carried heavily in our hearts. Nature truly does work in mysterious ways.

We couldn't wait to meet this precious new baby, but Mona, with her famously protective mothering instincts, didn't want us to get too close. We managed to get some good photos a few weeks later. We gazed in delight and admiration at this dear new calf.

We named her Sissi, short for Busisiwe, which means 'blessing' in Zulu. The name also referred to Sisi, the nickname of Elisabeth, Empress of Austria and Queen of Hungary. In choosing it, we acknowledged and thanked our wonderful guests and friends, Gudrun and Phillip Schneider, from Austria, who have supported Thula Thula's rhino conservation for so many years. Baby Sissi was the most adorable little creature and we all fell in love with her immediately.

But we weren't quite sure how this baby came to be. Had Thabo finally done the deed? Or was Mona pregnant when she came to us? Lynda and I got out a calendar and did the calculations. Mona arrived in May of 2019, and it was now March 2020.

'Not Thabo I'm afraid,' she said. 'She must have been newly pregnant when she arrived.'

'So the "boyfriend" I saw at Phinda must have been the father,' I said. 'I'm sorry it's not our boy Thabo, but who knows. Maybe next time!'

10

The Growing Family

Welcoming baby animals is one of the delights of the bush. In early summer, the impala drop their young, and the little buck are up and on their feet in minutes, wobbling alongside their mums. The baby zebras appear soon after, tiny fuzzy striped replicas. And the elephants. What a charming and delightful creature a baby elephant is, just three foot tall, with a crown of black hair on its forehead.

But big elephants are very efficient at making baby elephants, and those baby elephants spelled trouble for us.

When Lawrence passed away in 2012, we had twenty-one elephants, and were within the regulatory limits – but only just. The wildlife authorities told us that we had to take steps to control the population, and in September of that year, we implemented a contraception programme for our male elephants. Every six months, the bulls are darted from a helicopter. The hormone in the contraception stops them from getting sexually aroused (and, as a side effect, makes them generally calmer and more relaxed). This contraception programme is a good way to manage a herd's growth, and we were told that it should be reversible – when we stop the hormones, the elephants should be able to breed.

Contraception isn't a quick fix, however. The hormones take time to take effect, and a couple more babies were conceived before they did. We also discovered that a few females were already pregnant when we started the contraception programme – an elephant's gestation period is twenty-two months, after all, and you don't notice a pregnancy for many months, if at all. The result was that between 2012 and 2016, our elephant family increased to twenty-nine. Eight surprise births in four years. That wasn't the plan at all!

We had expanded into Fundimvelo in 2008, and in 2010, we had partnered with our neighbours, the Robarts family who owned the Lavoni farm to the south of us. The farm was 1,350 hectares, of which we incorporated the 1,000 hectares of bush. The remaining portion was partially sugarcane fields which had been leased to a sugar company until 2020.

All in all, we had three times as much land as when we started Thula Thula, with our first seven elephants. But despite the expansion and the contraception programme, we had reached our maximum carrying capacity for elephants. We needed more land for our existing herd and, hopefully, more elephants. Our longer-term goal was to allow our elephants to breed naturally. An elephant herd needs to have babies. All the females in the breeding herd look after the baby, and they all learn from that experience. Raising babies is an important aspect of elephant society.

Christiaan always says, 'The two happiest animals in the world are a farm dog and a baby elephant', and it's true. You only need to look at a precious baby elephant to see that! A calf gets 24/7 attention, and is cherished, nurtured, protected and instructed by the whole extended family. In fact, that's why orphan elephants are so difficult for humans to raise – we just can't replicate the constant love and attention of the herd. So yes, we needed more land so that we could abandon our contraception programme and have elephant babies!

It's not just about numbers. With elephants, you have to keep the family balanced and happy, with a good ratio of males to females, with babies coming up through the ranks, and some teenagers learning the ropes from the young adults. We had quite a few youngsters and teenagers now, but they were going to grow up. It was about time for another baby – the contraception programme worked, and we hadn't had any more surprises. More land would allow us to keep our beautiful herd in harmony.

Meanwhile, the authorities were becoming increasingly insistent that we develop a ten-year plan for our elephants, and that it comply with the latest regulations and protocols. It was non-negotiable – every reserve has to have a specific number of hectares per elephant, no matter what the terrain. Of course you need rules and regulations – you can't have people keeping wild animals in conditions where there's not enough food to eat or space to roam, and you can't have animals destroying habitat. But some land is able to support more game than other areas.

'Just look at our carrying capacity, compared to the Free State or even the Kruger Park,' Christiaan said, exasperated. 'We are almost part of the coastal belt. The bush is lush.

We have grass throughout the year. There's plenty to eat. We have hills and valleys, lots of trees . . .'

'I know, but it doesn't make any difference,' I said. 'They don't take the topography into account. We need more land if we want more elephants. We're just going to have to make a plan.'

Finally, a breakthrough! After three years of discussions, in December 2019, we reached agreement in principle with the *amakhosi*. A further 2,000 hectares of community land would be incorporated into Thula Thula. The relief I felt was immense. We would be compliant with the regulations, so that stress would disappear. And we would be one step closer to Lawrence's dream of a big nature reserve, a happy home for our elephants – and some more.

Land expansion is important for other reasons, too. The larger the area, the more diverse the topography and the habitat, and the more dynamic the ecosystem. If you have a bigger area and fewer fences, the animals can move about and you can have a greater density of animals without damage to the ecosystem. However, expansion isn't only a matter of finding more acreage. It means finding more money. When you expand, you have more land to protect and manage: new electric fencing, essential access roads, extra security guards and anti-poaching, clearing alien vegetation and invasive species, creating fire breaks. The list is endless, and all of it costs money. But I had learnt to try and tackle one thing at a time. First things first – the land. And we had it! All we needed was to finalize the paperwork and the next phase of the Thula Thula expansion programme could begin. It was all systems go. We would commence with the fencing in March 2020.

Well, everyone knows what happened in March 2020. The pandemic. And lockdown. All our plans were set back by the uncertainty and the lack of income. We put our fencing project on hold and focused on survival.

11

Covid Comes to Thula Thula

I welcomed the start of 2020 in an Ayurvedic resort in Kerala, India. After a hectic year I really needed a break and this was perfect – sea, sun, yoga, and no worries at all, far away from my insanely busy bush life at Thula Thula.

I try to visit India every year, and I particularly love Kerala – the beautiful tropical coastline, the culture, the food, the magnificent temples. But I'm always saddened to see how the elephants live. Elephants have been part of Indian culture and history since ancient times. They have been worshipped and used in war. The revered Hindu god Lord Ganesh is depicted with an elephant head and is believed to remove or overcome obstacles. Nonetheless, Indian elephants are often treated very badly. There are hundreds of domesticated elephants which are used in festivals and parades and in the tourism industry. Training is often extremely cruel, and they are kept shackled in chains. It pains me to see these magnificent animals dressed up in ornate cloths and bejewelled headgear, and draped with bells and necklaces, with people riding on them. When I see them lumbering on the hot tar roads under the sun, I think how different their lives are from the lives of our own elephants who are in a healthy, happy family herd, grazing and roaming and playing as they should.

It was while I was in India that I first heard mention of a mysterious virus in China. I wasn't particularly concerned. Another case of media fear-mongering, I thought. It would soon fade from the headlines like bird flu, Ebola and mad cow disease. By the time I left for South Africa in early February, a few cases of this virus had shown up in Kerala. But I was looking forward to getting home and starting the new year, and I was not going to be pulled down by this bad news. Besides, our little Zululand paradise was miles away from China and its troublesome virus.

I'm not one to sit glued to the TV news, but I could not help checking now and then to see what was happening. I was rather alarmed to see that people were dying in Lombardia in Italy. Schools were closed, said the news report, and cultural and public events cancelled. Within weeks, the word 'lock-down' was all over the news. First China, then Italy, then the whole of Europe kept their citizens at home. Thank goodness we were safely tucked away in beautiful Zululand where our only troubles were an unruly young rhino, the frustration of waiting for wildlife permits from slow officials, and those naughty vervet monkeys who loved to tease my dogs and set them yapping at dawn.

'The world has gone crazy because of a little virus,' I said to Lynda and Christiaan over coffee one morning. 'Closing all businesses for even a couple of weeks is going to damage the world economy terribly. Now just imagine if this was to reach South Africa!'

The first few cases were confirmed in South Africa in March – all people travelling back from Europe. Still, with my usual optimism, I believed that these were isolated incidents and that this trouble would pass us by.

'Well, I have been watching the news and it seems that this virus is spreading faster than we thought,' said Christiaan, who keeps up with the news more regularly than I do. 'With people flying in from all around the world, who knows . . .'

'It would be catastrophic for the South African economy,' said Lynda. 'Our government does not have the financial resources to take care of people in a lockdown.'

'It won't happen to us,' I said, ever the optimist. 'South Africa cannot be on total lockdown.'

Life in the bush is never smooth sailing – it's too unpredictable – but we had stood the test of time and, after twenty years of running this reserve and Lodge, February 2020 was our best month ever. The guests were coming from all over the world to enjoy our beautiful bush and meet our animals, especially our famous elephant family. Plus, we had the expansion to look forward to. I couldn't have been happier with our prospects for the year ahead.

On the evening of the 23 March, I was happily on the sofa, chatting to the doggies, giving them treats and sharing my happiness with them . . . when I heard the news. President Cyril Ramaphosa announced a three-week lockdown of South Africa to contain the spread of Covid. Schools and businesses were closed, travel was restricted, citizens were confined to their homes, and only emergency workers could move around freely. It was a devastating blow to the local tourism industry.

The Lodge and the Tented Camp were both full, and news of the lockdown created a total panic amongst the guests, particularly those from overseas. They knew they had to suddenly shorten their holidays and try to get onto the few remaining flights home. Information was scarce and contradictory. No one knew when the flights would be grounded, whether they would be allowed to enter their home countries, or whether they would have to quarantine and for how long. But they all started packing. There was a desperate, wartime feel to their departure, as if they were fleeing for their lives. There were no weapons, of course. No army, no generals. No outward sign of the risks we were facing. The fact that the enemy was an invisible virus made the situation feel quite surreal. There were many moments when I felt as if we were trapped in a bad science fiction movie.

On 26 March, Thula Thula closed its doors to guests. With travel restrictions coming into force the world over, international tourism came to a halt. Our income disappeared overnight. It was disastrous, but I knew we would survive.

'Come on everyone, it's only for three weeks!' I said. 'Don't despair.'

The empty Lodge felt very bleak without guests. The big glass doors to the veranda and lawn were closed, leaves floating in the swimming pool. The dining tables, usually covered in fresh white table cloths ready to receive the baskets of Chef Tom's homemade seed bread, were bare. No guests gathered at the bar to compare thrilling game sightings and swap stories about their bush adventures over a cold Castle Lager.

The buck took full advantage of the pandemic and the empty Lodge and moved in! There is an electric fence around the Lodge and its grounds, the wires set at a level that keeps out the bigger game and lets the smaller animals wander in and out. The nyalas and impala are often seen on the lawns of the Lodge, but now that things were so quiet and peaceful, the buck took up permanent residence, happily eating the grass and drinking from the pool, protected from predators. A herd of wildebeest soon joined them, wandering off into the reserve during the day to graze, and coming home to the Lodge to sleep in safety at night.

The Thula Thula elephants were less happy about the sudden quiet on the reserve. In fact, they were confused. Where is everyone? Where are the game rangers? The visitors with their cameras? Where is the afternoon game drive vehicle, full of admiring tourists? The elephants set out to investigate. In those first few days after lockdown, they visited the Lodge to see what was happening there. Not a lot. Then they headed down to Tented Camp – perhaps there they would find some human company? No such luck. They cruised past the volunteer academy, where there would surely be a group of admiring young people to welcome them? Empty. They swung past the main house to look for us, stopping by the fence and peering into the garden. Success! We all came out of our houses and offices to say hello and admire the elephant family gathered across the fence from us. We were as happy to see them as they were to see us!

'Gobisa! Marula! Ah, you look so well.'

'Look how Andile has grown!'

'Where's little Lolo? Oh, there he is.'

They stayed a while, the adults munching on the trees that hung over the fence, or wandering about, the little ones charming us with their antics.

'I think they missed us!' said Clément, my partner, who had come from Durban to stay at Thula when lockdown was announced.

And indeed, it did seem so. Elephants are very sociable animals, and the Thula Thula elephants, in particular, are accustomed to humans. They see the familiar green vehicles come out morning and afternoon, with the rangers they know well, and a few new people with their own unique voices, smells and emotional reactions to investigate. As much as we humans love to observe and interact with the elephants, they are observing us. They are curious about us, too. If our elephants didn't want to see the rangers and the guests, they would make themselves scarce, that's for sure. An elephant can hear a car coming miles away, and despite their size can disappear into the bush so quietly and thoroughly, that even if they were a few metres from you, you wouldn't see them.

Now here they were, reaching out to us.

'Not to worry, it's only for three weeks,' I said confidently to our elephant family. 'It won't be long now, your human friends will be back soon.'

As you know, that proved not to be the case. The lockdown was extended, and there were many challenges yet to come.

12

Making a Plan

As the matriarch of Thula Thula, an important part of my role in difficult times is to stay positive and keep up morale amongst the staff. We employ about fifty people at Thula Thula, plus our security and APU team. Our business relies entirely on guests, many of whom are international tourists, to keep the place afloat and to pay salaries. Now they were gone. Understandably, everyone was anxiously looking to me for guidance and wondering, 'What are we going to do?' I called an emergency meeting with my management team – Mabona, Christiaan, Siya, Lynda, Vusi, and Aphi, the office manager. Everyone looked nervous. I knew that they feared for their jobs, their staff, and the survival of their families.

I did not mince my words. 'There are difficult times ahead and we need to prepare ourselves. We need to make decisions for the survival of all of us, humans and animals,' I said. 'We will not stop paying salaries, so don't worry, you will be able to look after your families.'

'That's a big relief!' said Siya. 'So many other lodges and hotels are closing down or retrenching. We were all so worried.'

'This reserve isn't like other reserves. We're like a family. We can't abandon our staff and their loved ones,' I said.

'Thank you, Françoise,' said Mabona, her pretty face breaking into a smile.

Lynda went into practical mode. 'I'll apply right now for all the help we can get. There will be money coming from the Unemployment Insurance Fund to help with salaries . . .' she said. 'It's not much but . . .'

'It's a start,' I said. 'We also have some cash saved up for a rainy day or special projects.'

It was a tough decision to make. Dipping into the emergency fund meant we wouldn't be able to use that money towards our conservation plans, but these were desperate times. I knew I had to prioritize our staff. Most of them come from the villages surrounding the reserve, or a little further afield in rural Zululand. With the high unemployment in the rural areas, a staff member might be the only breadwinner in their family, supporting as many as a dozen people. If we retrenched, hundreds of people would go without food.

'We'll use the savings to fund salaries until I can think of something else. People have to eat,' I said. 'And this won't last. The lockdown is only for three weeks. The virus will disappear before long.'

Christiaan frowned. He had been very quiet and serious throughout the meeting. I knew that he was on the news channels every evening and did not share my optimism.

Nonetheless, I continued, with confidence, 'The Spanish flu lasted two years. That was a hundred years ago. Medicine has made such progress since then. Believe me, by May or June, everything will be back to normal. Don't worry, it's going to be alright!'

Eish . . . why was I getting it so wrong every time? My stupid eternal optimism had to stop now and I had to learn to face reality. It was not going to be alright. Christiaan's stony silence was him telling me, 'Wake up girl! This shit is real.'

'So, what are we going to do?' asked Aphi. 'There are no guests to look after.'

'I'll tell you one thing we are not going to do,' I replied. 'We're not going to give up and sit here watching television all day. We are going to focus on the positive and see how we can survive this and help others where we can.'

I allowed myself a moment of sadness, then I got to work on a plan of action. Staff were put on rotation. We divided them into three groups. One group was thoroughly screened at the clinic before coming to work for six weeks, during which time they stayed at Thula exclusively. At the end of the period, they would go home to their families in the villages, and the next group would take over. Everyone was put to work – the staff from the Lodge and from Tented Camp, all the rangers, even admin and reception, they were all kept busy. And they were all paid in full.

I believe that keeping your environment clean and well-maintained is essential for morale. And we had a big environment – the whole game reserve! There is never any shortage of work to be done – fixing fences, sweeping for snares, eradicating alien plants, clearing the bush along the roads so that vehicles would not get scratched. Maintenance is always a problem in a busy Lodge – when there are guests around, you can't be hammering and sanding, so when do you work on the place? Now was the time! Christiaan and Clément taught the ladies at the Lodge how to sand the wooden furniture and floors and then polish and varnish. Portia's boyfriend, a builder, stayed with us during lockdown and he was a great help with the maintenance tasks.

Fortunately, despite the lockdown, shops considered 'essential' were open – and that included hardware stores. They were able to sell us paints and brushes and varnish, but for reasons we could never understand, we couldn't buy nails. There were all kinds of silly new Covid-19 rules which did not make much sense. In the local department store, you could buy winter shoes, but not the sandals on the next aisle. The logic escaped me, but we all had to obey the rules like good little soldiers. The more immediate problem for many South Africans (including some of us – I won't mention any names!) was that the government banned the sale of alcohol, to take the pressure off hospitals, and cigarettes, for reasons no one could understand. A minister tried to explain that 'when people zol' (smoke a joint), and share it, they get saliva on it and spread the virus – well, she found herself memed all over the internet in minutes. People who wanted a tipple and a smoke were now criminals, buying contraband from smugglers in a flourishing new black market.

I looked over the deck one sunny afternoon. The chairs that are usually clustered around tables were upended, with the Lodge staff working away with their sandpaper. The sound of their Zulu chatter and laughter competed with the calls of the hadedas in the trees above.

'Look, everyone's enjoying it,' I said to Clément. 'People need to keep busy doing something useful.'

'It's true, *chérie*,' he answered, 'And see how good everything looks.'

He gestured to a few chairs, already varnished and set out to dry, gleaming like new.

Sne, our hostess at the Lodge, came out of the kitchen and called, 'Lunch is ready!'

The ladies put down their brushes and sandpaper. The rangers appeared, magically summoned by the smell of her cooking.

'I didn't even know Sne was a cook,' I said. 'You see? We have all learnt something from this difficult time.'

Sne smiled proudly as everyone tucked in, smacking their lips and praising her food. Showing appreciation is one of the simplest and most powerful acts in life. It makes someone feel seen and proud; it boosts confidence and makes us want to do even better.

Yes, in a funny way, life was good. We were tucked away in the bush, with lots of fresh air and few people around, so we didn't feel at any great risk ourselves. The future still looked uncertain, but we kept hope and, above all, busy.

My optimism had taken a bit of a knock when the twenty-one-day lockdown was extended. And every day or week that went by put us under more pressure. Four months into hard lockdown, I entered a period of deep discouragement. Meanwhile, infection figures were soaring, and there was concern about how the hospitals would cope. With each extension or bad news story, I felt more fear for the future.

The money was running out, and who knew when we would have paying guests again? I had no choice but to face reality. That 'little virus that began in China' was doing more damage than I could possibly have imagined. For the first time, I wondered if all those commentators and naysayers – the ones I dismissed as negative – were right. Maybe this pandemic would last two years.

The weight of my responsibilities felt overwhelming. How long would we be able to keep paying for our anti-poaching team, our security, our rhino monitors? What would happen to this beautiful haven and our precious animals if we couldn't take care of it all? Even with all this effort, we found cuts in the electrified fences. We had to constantly fix them and find and remove snares. Snares are almost invisible wire nooses that poachers use to catch animals. They are indiscriminate and terribly cruel. A poacher might lay down ten snares and catch one buck. When he comes back to check on the snares, he takes the buck and the snare, leaving the other nine snares in the bush where they eventually entrap other animals, leaving them to die a terrible death. There was no way we could stop our vigilance in the fight against poaching. I had made a commitment to protect these animals. And what about the staff? I had made a promise to them, too. I couldn't let them down, the humans or the animals. I knew I had to find a solution, something to help us all survive. But what? I racked my brains for ideas but felt quite helpless in the face of what was happening in the world. It was exhausting, and I couldn't help falling into moments of deep hopelessness.

Something that gave me strength to carry on was the

outpouring of love and concern from our friends around the world. They all knew that the hospitality and tourism sector was being badly hit, and they were concerned about Thula Thula, our animals and our people.

A few months into lockdown, Jo Malone phoned: 'How is it going there? How are you doing?'

'It's not easy, but we are surviving,' I told her.

I knew that Jo and her family must be battling too. The UK was in heavy lockdown. The shops were closed, morale was down. I found it extraordinary that even in such dark times, this beautiful lady found the heart to reach out with love and concern. During lockdown, nature provided us with an opportunity to honour our friend Jo in a unique way. Muzi came back from a game drive at Mkhulu Dam, his eyes shining with excitement.

'You'll never guess what I just saw!' he said. 'We've got a baby hippo!'

Hippos are large under any circumstances, and spend a lot of time submerged in water, plus they are most active at night, so it's hardly surprising that no one noticed that Juliet was expecting! She and Romeo produced the most perfect, darling little baby.

Baby hippos weigh in at twenty-five to fifty kilograms at birth, after an eight-month gestation period. It's not easy to determine whether the baby is male or female, but that was fine by me. We would name the little one Jo – a good unisex name – after our friend Jo Malone.

I called her the next day. 'Congratulations,' I said.

'On what?' she asked, surprised.

'You're the godmother to a beautiful chubby little baby hippo! We've named her Jo in your honour. She's waiting for you here at the Mkhulu Dam – you'll just have to come and visit soon.'

I sent her lots of photos and videos of her new godson (yes, Jo turned out to be a boy!), who we called Baby Jo. I ended each message with the little red heart emoji for 'love', as her brand was called Jo Loves. It was as if Baby Jo was her little African mascot or good luck charm.

A birth is always a gift, all the more so in the hardship of lockdown, and all the more so when the species conservation status is listed as vulnerable. Yes, hippos are on the decline across Africa because of hippo–human conflict and unregulated hunting for their meat, skin and teeth. Romeo and Juliet have done their bit, I have to say, popping out babies Chomp, Chocolat and now Jo.

'Romeo could give Thabo a few pointers in that department,' Victor said with a laugh. 'Teach him some tricks.'

In difficult times, I had a moment of pure gratitude and joy – a precious new life, a baby hippo who would always remind me of the love and generosity of a dear friend. What extraordinary gifts.

13

The Dogs That Rule My Life

I am a positive person by nature, and I honestly believe that adversity opens the way to new opportunities. Although I was worried about how we would pay the bills, I was already looking ahead towards the reopening, and making sure Thula Thula had guests lined up. I knew we couldn't stop marketing.

Kim had left Thula Thula to travel and was doing photographic work in Limpopo province at the time. She was just the woman for the job! I called her. 'Come and stay,' I said. 'Bring your cameras, we can make some beautiful films and take pictures in lockdown. You can help with our social media, show off the reserve and our wonderful animals, so when the lockdown is lifted and things get better, we'll have lots of bookings.'

Kim jumped in her car and drove to Thula Thula.

The volunteer academy, where Kim stayed, was created to educate and inspire people about conservation. From the moment we opened in June 2018, it was a great success with young (and not so young) men and women from all around the world. They can come for a weekend, or a week or two, or for a full twenty-one-day programme, and their days are shared between working on the game reserve, and an educational programme in the afternoon. The volunteers pay a small amount to participate, and these funds are used for our conservation projects.

Through the volunteer academy, we also host school groups, and we do outreach in our local communities. I want people to dream about the bush, to fall in love with something about it, to care about it. I believe that if you want people to care about conservation, and understand what is at stake, they have to really experience the bush in all its beauty and hardship. When you have been involved in clearing bush by hand, or have seen a little buck struggling in agony in a poacher's snare, or been up close to an elephant and looked into its wise and gentle eyes, you gain a better understanding of the 'behind the scenes' of running a game reserve and the efforts deployed to protect our wildlife. Every one of our volunteers has an extraordinary experience. Even if they don't go into conservation as a career, they might grow up to be a politician, or the CEO of a bank, or someone else who can make a difference.

The volunteers live in a rustic permanent tented camp on top of the hill, with the most beautiful view over Mkhulu Dam and the rest of the reserve. There is also an old house and a rustic outdoor kitchen on a hill under a giant flamboyant tree. Acacia trees are dotted about for shade. Aloes produce bright nectar-rich flowers in red, orange and yellow in the winter months, to the delight of bees and butterflies and an incredible array of chirping and chattering birds. You can see the sun rise in the morning and set in the evening, and in between you can watch the elephants, rhino and giraffes make their way across the plains, from dam to forest to grassland. It is a blissful spot, but it is far from the main house where I live with our resident pack of dogs. With no volunteers around, Kim found it rather isolated and lonely.

'I need a dog to keep me company. Can I take one of them home with me?' she said, pointing to the dogs scattered about on the carpet and the furniture. 'What about little Lucy, can I have her?'

Lucy was an adorable little brown mongrel who had appeared at the gate a few months before. The guard called to say there was a puppy hanging around, and Andrew went to check. He arrived back with her – skinny, starving, shaking and full of fleas. Lynda bathed her, fed her and took care of her. She was our new baby and we loved her.

So when I saw Kim eyeing her for her new pet, I said, 'Not a chance! You're not having my dogs. Just go get your own, OK?'

Well, Kim took that literally and was back thirty minutes later with a tiny, shivering dog. She had seen some village children with three puppies, just skin and bone, and asked if she could have one. She promised she'd look after it and give it a good home. They said OK, and she picked a brown female with a black muzzle and soulful eyes.

'Here she is,' said Kim, putting the little thing onto my lap. 'I knew if I took too long you'd change your mind, so I went immediately! I am calling her Zara.'

'Welcome to paradise, my little Zara,' I said. 'For sure, I am going to fatten you up, skinny girl!'

Fattening up dogs is one of my great talents – I just can't resist their pleading eyes, and within months of joining the family, most of them are a little, shall we say, voluptuous.

Kim picked her up and proudly showed this poor little creature to the rest of the office staff. 'Isn't she cute? I think she's just beautiful.'

The ladies in the office made a big fuss of her. Andrew's gruff response was, 'Don't leave her out on the lawn on her own. One of the eagles will take her.' At that, Kim tucked the dog into her jacket, where she spent most of the day in her first weeks with us.

Zara was so weak and thin, we weren't even sure she would survive. We had her inoculated and checked by the vet and tried to build up her strength and her trust. She was scared of her own shadow. We never understood why, but she wouldn't walk through a doorway or over a threshold.

'Come on Zara, you can do it,' Kim called from outside. Zara stayed put, looking pleadingly at her.

Kim waved a biscuit. 'Snacks! Come on.'

Zara hovered nervously at the doorway but wouldn't go out.

'Oh alright,' said Kim, scooping her up and carrying her through the doorway.

Zara walks through doorways now. She is very happy and confident, although she still likes to keep close to Kim. Zara loves a warm spot, and if you sit on the sofa, she will snuggle in behind you and gradually push you out of the way until she's lying on the nice warm cushion where you were sitting and you are perched on the edge of the cold sofa.

I am pleased that we could give this scrap of a dog such a happy life here with us. I am a passionate believer in 'adopt don't shop' when it comes to dogs. All our dogs were abandoned, sick or unwanted when they came to us. I love dogs. It is terrible to think of these loyal and loving animals being mistreated or neglected. But it makes me happy that in spite of their difficult starts in life, our Thula Thula dogs have ended up in doggie paradise.

Now let me introduce you to our little Mauritian princess, the beautiful honey blonde Tina.

In 2019, soon after we arrived for a three-week holiday in Tamarin, in the south-east of Mauritius, this skinny, mangey little street dog appeared on the veranda of Clément's family house where we were staying. Of course, I fed her – she tucked into a bowl of free range chicken and basmati rice with great gusto – so it was no surprise that she settled in. She stayed and appointed herself our guard dog, taking up residence on the veranda and barking at passers-by. She was so cute, so kind and affectionate. As the end of our holiday approached, I knew I couldn't go to the airport and leave her behind.

On our last weekend in Mauritius, I said to Clément, '*Chéri,* I love this little dog already. We are going to take her back to South Africa.'

'OK,' he said. He knew me too well to try and dissuade me. 'But how are we going to do that?'

Our flight was on Wednesday, so we only had two days to sort out the inoculation, the kennels and the quarantine. First thing Monday morning I got to work on arranging

Tina's immigration, starting with a phone call to Dr Bester, a local vet who had been recommended to me. I introduced myself and started to tell him the story about my little dog.

'Wait,' he said, 'Françoise Malby-Anthony? Are you the one who wrote the book about the elephant? I've just read it. Come now!'

I went straight there with my beautiful little Tina.

'Have you ever seen such a cute dog?' I asked proudly. 'What do you call this sort of dog in Mauritius?'

He looked at the little scrap and said dismissively, 'We call them Maurichien.'

It was a pun on the word Mauritian – *Maurice*, French for Mauritius, and *chien* meaning 'dog'. I did not take that as a compliment! Basically, it means she's no breed at all – she's just a dog, a nobody, a regular Mauritian street hound. That may be so, but she left there with a full deck of vaccinations, all the blood tests, and a nice fancy collar and lead. I engaged a pet travel company to arrange the details. Poor Tina, she looked so confused to find herself shut up behind bars – in a five-star kennel, mind. I put my sarong in her cage, hoping that she would be comforted by the smell of me, and left, feeling very sad.

Two months later, on 30 January 2019, Tina arrived in Durban. She recognized me straight away and ran towards me wagging her tail and looking for pats and cuddles. We were both so delighted to be reunited! She is still a very happy, good-natured and affectionate dog, who seems to know that she got very, very lucky.

When we fetched her, I thought about all the trouble we

had to go to to get a dog into South Africa, and I said to Clément, 'What a performance. Imagine if humans had to get all those tests before we travelled anywhere.' Little did I know that a year later, a virus would arrive that would require just that.

Like me, Lynda is a crazy dog-lover with a soft heart. She hadn't been here long when our local animal shelter called to say they had a mother Labrador with seven puppies, some golden and some black. We went to take a look. The mother didn't look well, and we wondered how she would feed those puppies, and whether they would all survive. I said to Lynda, 'If you want one, get one, and I'll get one. We can't save them all, but we can save two.'

Lynda, being a clever farm girl, chose the biggest, strongest one, a golden Lab who she called Miley. I got a little black one and named her Chéri. Both puppies soon became very sick with parvovirus. Miley survived, but poor Chéri died within a week. We later discovered that the rest of the litter and the mother got very ill and had to be put down. Our Miley is the sole survivor, a big, friendly girl and a typical Labrador. She might be fast asleep at the other end of the garden, but if I open the fridge, she'll be right next to me with her big brown eyes, pleading starvation and hoping for a treat from the big white box from which delicious snacks seem to emerge. If you ever meet Miley, you will see that she is certainly not being starved!

Thula Thula might be famous for its elephants, but it's the dogs that rule my life. Every one of them has their own cushion or bed. Even so, the sofas in my house are covered with blankets, because our dogs sleep, sit, walk and jump on the furniture. I cook them special soups and stews – free-range chicken, mind you, no battery hens for me or my doggies – and I keep a big glass jar of dog treats on my desk just in case anyone needs a snack. I am always letting a dog out or letting a dog in, cooking for, feeding, patting or talking to them.

I often have eight dogs in my bedroom, which doesn't make for the most peaceful night as they take their chances trying to jump up on the bed. Inevitably, someone will start barking at a buck in the garden or an imaginary intruder and then they'll all join in in a great cacophony. But this is what we signed up for when we decided to adopt these little inno-cent and abandoned creatures. We are all part of the pack, for better or for worse. And I have no regrets – even after a sleepless night.

Every time I leave home to go down to the Lodge or the Tented Camp, the dogs look at me with sadness in their eyes, as if they fear being abandoned. And when I come back to the house, some half an hour later, they welcome me as if I had been away for days. Witnessing this unconditional love, I just tell them softly, 'No my little angels. You know I will never leave you.'

The dogs benefit so much from the communal lifestyle here at Thula Thula. They move between the houses in the compound at main house, or visit the ladies in the office. If I go away, my dogs just move in with Lynda or Kim. I look after their dogs when they go on leave. There's always someone to give a dog a treat or a pat.

I wish that the animal shelters could be empty of the abandoned creatures they care for, and that every dog in the world could be as happy and spoiled and loved as the dogs of Thula Thula.

14

Thabo vs The Big Digger

As well as the sanding, scraping, painting and polishing, there was one huge maintenance job that needed doing during lockdown – fixing our dirt roads. They were in shocking shape and getting worse by the year. There were potholes and dongas, teeth-rattling ridges and rocks, and slippery sandy bits that were quite treacherous when wet. After a hard rain, we couldn't drive on some of them for hours or even days – it was just too dangerous. The roads were empty of guests and game drives, and it would be the perfect time to work on them. But fixing up roads isn't like sanding furniture. It is a big and expensive job. And besides, surely guests would be coming back soon?

As summer came to an end, and there was no sign of the easing of travel restrictions, I decided to bite the bullet. I would dig into our savings and take the opportunity that Covid presented me with. The dry winter months stretched ahead of us, and there were no guests and no game drives. There would never be a better time to fix the roads. In May 2020, we got to work on this mammoth task. We used the enormous excavator, a TLB, to dig quarries for the stone and gravel we needed.

Thabo hated that excavator. Rhinoceroses look big and

tough, but in some ways they are actually very sensitive creatures. They have some of the keenest hearing in the animal kingdom, and can hear sounds and frequencies that we humans can't. They also pick up vibrations through the ground. The rumbling, digging, grinding and crashing of the earth mover drove Thabo mad. Imagine your next door neighbours having a week-long all-day, all-night dance party, with maximum bass, whilst simultaneously using the leaf-blower in the garden – that's how poor Thabo felt!

Vusi came to me halfway through the road building process and said, 'We have to do something about Thabo. Look at this.'

It was a video of the TLB on a mound of sand, with Thabo standing with his feet on the top of the mound and his big head looking into the cab. He seemed to be having a stern word with the driver. I imagined their conversation.

Thabo: 'Who are you and what do you think you're doing making all this racket on a Sunday morning?'

Driver, waving his arms in a panic: '*Eish!* What the hell? Get away from my excavator, big fellow!'

Thabo: 'How about YOU move away, unless you want me to push you away. This is my territory, little human. Now stop making that horrible noise!'

The driver, realizing he was dealing with a very unusual and determined rhino, turned off his engine.

Thabo: 'That's better. Now please don't make me come back and ask you again!'

And with that, Thabo turned on his heel and trotted off, leaving the poor driver shaken but unharmed.

Thabo hates diggers, tractors, rollers, trucks – but he has a particular loathing for the sound of a chainsaw. And honestly, who can blame him? As part of our lockdown maintenance plan, we had a team clearing sickle bush, an aggressive plant which quickly becomes too thick and takes over the grassland. We wanted to get rid of a big area of it to create a more savannah-like terrain in some areas. Hence the chainsaw.

The endless irritating noise got Thabo's attention. He went in search of the source of this racket, perhaps to have a word with its owner. After all, he had successfully shut up that noisy TLB. What he didn't know is that we were prepared for his arrival. We couldn't risk having men on the ground with noisy tools getting a visit from our big boy. Wherever they worked we stationed a whistleblower up a tree. He kept a sharp eye out and if Thabo was spotted – as he often was, attracted by the noise – the whistle blew and the men hot-footed it to the safety of a nearby vehicle. Thabo must have thought he had magical powers. He arrived to investigate the dastardly noise and, poof, it stopped! Just like that. Well done, Thabo!

On one occasion, the whistleblower didn't get down from the tree in time to join the rest of the team in the car. Thabo made a slow circle round the tree and then looked up, surveying this strange arboreal human with great interest. It was nice and quiet by now – thanks to Thabo's magical powers. What better time for a nap? That poor man was clinging to the tree for an hour while Thabo dozed, before he finally trotted off, refreshed, to see if any other noisy machines needed to be exterminated.

'The guys are scared,' said Vusi, and honestly, who wouldn't be, with a big rhino appearing at their vehicle's window or under their tree? 'If we don't do something about Thabo, they're going to quit. And besides, the noise is going to drive Thabo mad!'

I could see his point. We needed to keep Thabo and the workers away from each other, for both their sakes. But what to do?

'He'll have to go in the naughty corner,' I said. 'We need apples!'

A plan was hatched to entice Thabo to the *boma*, lock him in, and keep him there while we finished the road mainte-nance. He would have peace and quiet, and the workers could go about their business without being disturbed by our little hooligan. I bought a big bag of apples – his favourite treat from when he was little. The idea was that Promise would drive the vehicle and Khaya would toss out the apples. Thabo would follow the trail of apples all the way to the *boma*, just as Hansel and Gretel followed the breadcrumb trail through the forest in the fairy tale. Kim would document the whole adventure in photos and videos. Brilliant plan.

But first, they had to find him.

They set off bright and early. I stayed home and waited for news. Regular WhatsApp messages from Kim kept me updated on their progress:

We found Thabo!

OK, he's eating an apple

Slowed down. Doesn't want the apples

I think he suspects it's a trick

This went on for hours, slowly, slowly enticing Thabo across the reserve to the *boma*. He lost interest in the apples, and Khaya had to lure him in by pretending to put something down on the ground behind the vehicle. Thabo was suspicious but couldn't resist coming for a look. He's a curious chap and he loves to see what the humans are up to. Metre by metre, they edged towards the *boma*.

And then I got a message:

Thabo has blocked us inside the boma and we can't get out!

I didn't know whether to laugh or rush over there in a panic with reinforcements. What had happened, I discovered, is that they finally got Thabo right up to the open gates of the *boma*. Promise reversed the vehicle slowly into the enclosure, with Khaya still tempting and cajoling Thabo to follow them in. The plan was that when Thabo was inside, they would scoot round him and out of the gate, shutting it behind them.

Except that Thabo hadn't approved the plan. In fact, Thabo had ideas of his own. Once the vehicle was inside the *boma*, he lay across the open gate and settled down for a snooze. There he stayed for the next hour. Promise and Khaya and Kim sat in the open vehicle in the blazing sun while Thabo had a nice nap after the day's exertions. Kim got lots of pictures of a sleeping rhino. Two hours later, the three of them arrived home tired and hungry, but with their mission accomplished. It had taken them the whole day to get that wily boy into the *boma* – and themselves out.

Khaya described how, when Thabo finally woke up, they

started the engine and reversed, gunning the engine to annoy him a little, so he would come in after them. It was a risky plan – as a general rule, you don't want to annoy a rhino! – but it worked. Thabo trotted inside. They managed to manoeuvre around him and out the gate. Khaya jumped down, slammed and locked the gates, and the job was done. Thabo inside the *boma*, the rangers and Kim outside, the excavator driver and bush-clearers safe. Everyone happy.

'Thabo is one smart rhino,' said Kim with grudging admiration. 'Never a dull moment with him around!'

15

We All Depend on Each Other

Thula Thula is in the north of KwaZulu-Natal province, in a rural area of Zululand dotted with small villages. It's a beautiful place, with rolling hills, brightly painted little houses and the round, thatched 'beehive' huts traditional to the Zulu people, but it is very poor and with few jobs or government services. As the lockdown took hold, and the weeks and then months passed, the people in the villages suffered terribly.

As well as the humanitarian crisis, this was a conservation crisis. Larry came to me with worrying news from the security and anti-poaching team.

'Our foot patrol guys are seeing more activity, Françoise,' he said. 'We're finding breaches of the fence, and more snares . . . It looks like people are poaching for food, or to sell the bush meat for cash.'

'They are only just getting by in the best of times,' I said, 'but with no tourists coming and jobs being cut during Covid, people are literally starving. We need to do something to help them.'

Help came from far away on the other side of the world.

We are very lucky to have loyal, supportive guests – or I should say dear friends – who have a long and deep relationship with Thula Thula and have taken us into their hearts. Some, like the Simonsen family from Denmark, have been staying with us for three generations. They have good relationships with the staff at Thula Thula and support our conservation projects most generously. So I wasn't entirely surprised when I heard that Susanne Simonsen had been in touch with Mabona and Promise to see how they were doing, and how their communities were faring in these difficult times.

'I told her that people in Buchanana have no work, and families are trying to survive on a small government grant of a few hundred rand,' said Mabona, who comes from that village. She has a very special relationship with the Simonsen family – they had invited her to spend her holiday in Denmark the previous year, showing her a marvellous time.

'Susanne says the family wants to help,' said Promise. 'They want to give us money to buy food parcels for the people who are suffering.'

I felt tears come to my eyes. It was extraordinary to me that in these times of great need and great difficulty the world over, this kind woman had thought about our little village in Zululand and reached out to see how she could help.

Before the week was out a generous donation had arrived from the Simonsen Foundation. It went straight into the Conservation Fund, Thula Thula's non-profit organization, to be used to feed families in the surrounding area. We immediately set to work to put the money to good use in the community.

'I want to give them nice food and some treats, not just

the most basic necessities,' I said to Victor and Promise, our volunteer academy leaders, who were in charge of buying the food for the parcels. 'And each parcel must be enough to feed a family for a month.'

They organized food parcels from a big supermarket in Empangeni, our nearest town, loading up on cooking oil, long-life milk, orange squash, bulk bags of chicken pieces, samp and beans, sacks of rice, and more. We packed it all into hampers. It was a big job, but it was very pleasing to see those big food parcels piled up, ready to go out to our community. The five *amakhosi* identified families who required help in their tribal areas, and we set out to find the needy families. Some of the little homesteads are so remote that they are not even on a road, and we had to carry the food through the bush or up a little footpath. Sometimes the family would spot us drawing up on the main road and come running down a dirt track with a wheelbarrow to collect their provisions. It took days to make the deliveries. And what joy when we arrived! They couldn't believe their eyes when they saw the bakkies arriving with the food parcels.

Little children came running out of their huts to welcome us, calling, '*Sawubona* . . . Hello' and clapping and laughing.

The ladies waved their hands and ululated, sending the high-pitched, trilling '*lalalalalala* . . .' sound ringing across the rolling hills of Zululand.

Many hands arrived to help unload the bounty. The men and women happily shouldered the sacks of rice and maize, the children running after them, carrying bottles and bags. The response was incredible, and so heartwarming. They live with so little – and with even less since Covid – that getting a bag of food was like winning the lotto.

'*Ngiyabonga* . . . Thank you . . . *Ngiyajabula* . . . We are happy . . .' they shouted, waving us on our way.

The people in the villages have a lot to teach us about life. They will share what little they have with a neighbour who has less. In spite of living in poor circumstances, they still sing and dance, and find moments of happiness and joy in difficult times. This work in the community was an extremely emotional experience and quite an eye-opener for many of us. Victor came back quite shaken after coming across a man living alone in a makeshift shelter in the bush.

'He had nothing,' he told me. 'Absolutely nothing.'

'A blanket, surely?'

It was winter and the nights were very cold.

'No bed, no blanket, nothing. What can we do for him?'

Mabona the Lodge manager collected blankets and towels from our stores, and even managed to find him a mattress. Victor delivered them, along with some food. We all felt better knowing that he would at least sleep warmly that night, with something in his tummy.

As much as the smiles and happiness brought me joy, it also brought tears. Most of the people in the villages had little or no education, and most had no job. If they do have work, it is probably in Richards Bay or Empangeni, a long and expensive trip that would eat up most of a miserable salary. Their prospects for a better life were slim.

I thought of my staff at Thula Thula, young ladies and men who were being trained and educated, had a roof over their heads and earned a decent salary. I was proud of them and wished I could employ more of the villagers. I couldn't help but feel disgusted and furious at the lack of care from the authorities, and the corruption that contributes to the extreme poverty in South Africa. The disparity between these rural people who don't have even the barest minimum to eat, and the people driving Bentleys and Ferraris and living in mansions in Cape Town or Johannesburg, is sickening. I tried not to dwell on that though. I would rather we put our energy into making a difference where we can.

And we did make a difference, but I knew that there was a bottomless need, and those food parcels wouldn't last forever. More friends came to the rescue – this time Kingsley Holgate and Grant Fowlds.

Kingsley is a well-known humanitarian, adventurer and writer. The Kingsley Holgate Foundation is involved in conservation, and in the fight against malaria, a disease that kills hundreds of thousands of people in Africa every year, many of them young children. Education, the distribution of insect-repellent mosquito nets, and indoor residential spraying are used to prevent the illness. Even in the pandemic, we can't afford to neglect important services like basic healthcare, and malaria prevention.

Grant is with Project Rhino, a non-profit organization concerned with saving rhinos in the province of KwaZulu-Natal. They bring conservation and anti-poaching messages to the wildlife communities and engage local children through art and education around rhinos, through the Rhino Art programme, run by Richard Mabanga.

Recognizing the growing food crisis in poor communities near game reserves, they turned their attention to helping them. In the first year of Covid, they gave out a million meals of their specially designed pre-cooked nutritional porridge. Grant saw our food donation efforts on Facebook and phoned with his usual cheery can-do manner.

'Hey Françoise, we want to come and distribute food parcels to your community. How's next Thursday?'

'That's wonderful! We will all help with the distribution.'

'It's going to be a long day. We'll have to stay the night. I'll bring tents.'

'We do have tents, you know Grant. The whole Tented Camp! It's empty, as we haven't had guests in months. We'd love to host you.' And then I added the words I knew would cheer the man's heart in these times of prohibition, when alcohol wasn't for sale: 'And we've got beers!'

Twelve of them arrived, the most wonderful people. Grant had a knack for bringing together interesting, caring individuals, many of them well known sportsmen or artists, to support his causes.

With his white hair and bushy white beard, Kingsley looked like a bush-whacking version of Father Christmas, handing out the bags of porridge, each enough for one hundred meals.

'You don't need to cook it,' he explained, giving a five-kilogram bag to a local *gogo* (grandmother). 'Just mix with water or milk. It has all the nutrients, everything you need.'

Grant was chatting away to the elders in fluent Zulu, asking them about their lives and the impact of Covid. Richard regaled the local children with facts and stories about rhinos.

After a long day on the road, we were tired, but buoyed by the warm welcome and goodwill we had experienced. We returned to Thula Thula for an evening of classic bush hospitality. Over dinner in the *boma*, around a roaring fire, the stories – and yes, the beers – flowed freely. David Jenkins, a young singer originally from Empangeni, was with us. He had written a song, with lyrics by my friend Jos Robson, about Thula Thula and the elephants that Lawrence had saved twenty years ago. He sang that beautiful song that evening around the fire.

I sang the chorus quietly to myself, savouring the beautiful words:

> *Thula thula*
> *Wo thulani*
> *(Calm down)*
> *Ningasabi sizoniphatha kahle*
> *(Don't be afraid, we will look after you)*

These evenings together are always characterized by wild bush tales – a giant cobra in the cupboard, two black rhinos in a territorial battle, many vehicles getting stuck in sand or mud or rivers. I must say, you can't ask for better dinner companions than conservation people and African adventurers. As the night darkened and the brilliant stars came out, the hyenas echoed our own laughter. It was a fitting end to a good day, and a truly magical evening in the midst of this Covid nightmare.

I feel very proud of the work we did to help our

community survive the pandemic and the lockdown. Although our own situation was terribly difficult, we were still able to reach out and show love and care to others. This is the way it should be. Caring, giving, helping, sharing. Not looking away thinking that someone else will do it.

This is what you learn when you live with elephants. Compassion. If one of them is in trouble, the whole herd will come to its rescue. I remember a story told by a man who travelled to Botswana. Observing a herd of elephants at the waterhole, he noticed that one had lost two thirds of his trunk. An injury like this would mean death to the elephant. With a shortened trunk, he wouldn't be able to feed himself, or get water. He watched as other members of the herd pulled water into their trunks and emptied it into the mouth of the wounded elephant. Now isn't that the way things should be in this world?

In nature, and in life, we all depend on each other.

16

Desperate Times, Desperate Measures

Week by week, month by month, the pandemic continued, and Thula Thula remained closed to guests. We had cut expenses to the bone, but on a reserve, there are many essential expenses you cannot cut. Our responsibility to protect and conserve this wildlife doesn't disappear because of a pandemic. Even if we have not one guest, we have to maintain our fences and pay for security. The poachers don't obey the government's 'remain in place' orders; if anything they are more desperate, more dangerous.

We have faced floods, bush fires, poaching, attacks on our animals and our staff, the death of our founder, my husband, not to mention the everyday challenges of living in the bush. In twenty years of running the oldest private game reserve in KwaZulu-Natal, we had never had it so tough. We had received a little money from the government's Temporary Employment Relief Scheme to supplement salaries. In September of 2020, that had stopped. We were on our own.

The responsibility of keeping the reserve going, paying the staff and keeping the animals safe weighed heavily on me. I maintained a happy face and a positive attitude in public, and spent my moments of discouragement and despair alone, with my dogs. They are the ones I turn to in difficult times, the ones with whom I can share my deepest fears and worries. Animals feel your sorrows and they are the best companions in times of distress. Gypsy will climb onto my lap to comfort me, and the others gather round. I felt loved and supported, and encouraged to go on, to keep up the fight – for their sakes, as well as my own. It cheered me up just watching these little survivors living their lives without a care in the world, except perhaps wondering what their evening meal was going to be, and when would be the best time for their pre-nap nap. I took heart from that. We were going to survive this crisis too! I just needed to find a way.

It was absolutely imperative to find ways to generate money. We have wonderful supportive guests and friends around the world. But I also knew that everyone was under pressure and donor fatigue was setting in. To just say 'Please give money' is very crass and boring. I think fundraising for conservation should always be for something specific, and I always want to give donors something of value in return. But what?

We were already running a successful virtual adoption programme. Our elephants, every one of them from biggest to smallest, were up for adoption. And the rhinos, Thabo, Nthombi, Mona and baby Sissi. Who can resist the adorable Sissi? We promoted our adoptions, and took gorgeous photos of the animals, posted them on our social media accounts, and shared them in our newsletter. Our adoption programme had been a success. So, why stop with the wildlife? I thought. We put our beautiful dogs up for virtual adoption, too. Our lovely rescue doggies looked too beautiful in their photos, and our friends and supporters responded eagerly.

I do believe that challenge and adversity force us to come up with new ideas and new ways of doing things. We had challenge and adversity, that's for sure. Now we needed the ideas! And I had one.

I called everyone together and told them.

'These are desperate times. We need to do whatever we can to raise money and survive until this situation improves.'

Everyone nodded in agreement.

'So what do we have? We've got these good-looking rangers; they are very popular. Khaya, Siya, Victor, Muzi, Andrew. All of them, we will put them on the market.'

There was not so much nodding at that. The rangers looked at me warily. Oh dear, they thought, the stress has got to Françoise, she's gone a little bit crazy!

'Seriously, it's a great idea!' I said with conviction. 'You know how the guests love our rangers. They are with them for hours a day on game drives, they get to know them, they learn so much from them. I'm sure they'd love to support them. Why don't we put them up for adoption, just like our special wildlife?'

It was decided. We would offer the rangers for virtual 'adoption' at fifty dollars a month for as long or short a time as donors want. The money would go to our Conservation Fund, and every cent of it used for the upkeep and security of the reserve and the animals. Kim took beautiful, glamorous pictures of the rangers and told a little story sharing info about each one – Muzi's wonderful singing voice, Victor's photographic talent, Khaya's great sense of humour, Siya's amazing knowledge about wildlife. She even got a smile out of Andrew, who is not what you might call an eager photographic subject.

What made it so successful is that everyone who has been to Thula or read my book knows and loves our rangers. And here's the clever part of the offer – if you adopt one of our rangers, you get more stories and photos! Our rangers really are very special. Unlike other game reserves, where game rangers stay a year or two and then move, most of ours have been here for ten or fifteen years. Regular guests get to know them, and often ask for their favourite to take them on a game drive. The rangers know the guests too – as Muzi says, you have to learn to read the people, not just the animals! Most are from Zululand, and were 'brought up' here at Thula Thula, which is to say they became professional rangers here. Siya, Victor, Muzi and Khaya all started in security, and came to love the bush and the animals.

While we compiled the rangers' stories for our adoption programme, they chatted about their early days in the bush. Khaya recalled how he was so taken by the experience of being in the bush that on his free days, instead of going home, he would often stay on the reserve and go out on game drives with the rangers, listening and learning from them. He knew then that he wanted to join their ranks.

The first step towards becoming a ranger is to be a tracker – that's the guy who sits on the seat on the front of the car, looking for footprints and dung and other signs in the sand, to show where the animals are. It's a scary job to start with.

'The animals come so close to you. And if an elephant or rhino pushes or chases the car and the ranger reverses, you are right there in front of it, with it coming at you! I was terrified,' said Khaya with a laugh. 'But I soon started to relax. Before long, I wasn't even scared of the elephants.'

'What's to be scared of? It's not as if they are dangerous – like, for example, frogs!' said Muzi, who is warm and full of fun, always quick with a joke. His comment raised a laugh because everyone knows that Khaya is scared of frogs. As a result, he has on more than one occasion put his hand in his pocket to find that a frog has been deposited there.

'I can't help it, that's how I was brought up!' said Khaya with a smile and a shrug. 'The adults said if you play with a frog or a chameleon, it will bring lightning down on you.' He gave a shiver at the thought.

'But the worst is crabs. When we were kids, our parents would say "don't play with crabs, if one pinches you, if you are a boy you will become a girl, if you are girl you will become a boy". Of course, no one wanted that to happen.'

'Does that sound likely?'

'Of course not. And I didn't know anyone who died of frogs and lightning either. Now, I think of it as a conservation message, a way to make sure we kids didn't mess with the small animals and hurt them.'

Today it was Khaya's turn to be teased, but everyone gets a turn to be the butt of the jokes. The rangers love to play tricks on each other, and especially on the new recruits. They swap war stories of their own.

'Siya told me that you can tell whether an elephant poo

comes from male or female,' Muzi recalled. 'You just stick your finger in the fresh pile of dung and put it in your mouth. If it's sweet, it's from a female and if not, male.'

'You did that?' Victor laughed.

'He did it first, so I thought, well that's what everyone does, so I must do it. The trick is, after he stuck his finger in the dung, he swapped fingers. He sucked a clean finger. I didn't, so I got a taste of elephant poo!'

Now they were on a roll.

'What about the trick with the acacia tree?'

'I still have scars!'

The beautiful acacia trees with the wide round crown that are typical of the African bushveld have long, sharp thorns. Luckily for the giraffes and other herbivores, the thorns point downwards, so when the animal wraps its tongue around the branch and pulls it down to remove the leaves, the thorns are flattened against the branch and don't hurt the animal. Rangers like to demonstrate this fact by clasping the branch in hand and pulling sharply over the thorns to remove the leaves. No injury! Except that there is one type of acacia which doesn't have the thorns angled downwards. The ranger will stop at one of them and ask the new recruit to demonstrate how a giraffe feeds on the acacia trees. The poor guy gets a handful of thorns.

As long as everything works out in the end, someone else's misadventures are cause for hilarity. One rainy night, Muzi was doing a fence check on a motorbike, when he found himself surrounded by the elephants. Now, this was a potentially dangerous situation. Elephants don't like loud mechanical noises like drones, motorbikes and helicopters, and when they come across them, they often chase or charge them.

'I saw the tusks shining at night. I was right in the middle of the herd so I stopped and turned off the engine,' Muzi said, when he described his adventures to the rest of us. 'I had to go past. I couldn't stay there. My heart was racing. What must I do?'

We all listened as he described his exit strategy. He was on a downhill slope, so he turned off the headlights, rolled a little way past the elephants, kick-started the bike and raced downhill.

'I made it! I was so happy to be out of that dangerous situation. I put my lights on and started for home. The next thing I felt a huge crash that knocked me right off my bike. A big male impala, blinded by the headlights, had run straight into me. I lay there in the dark with stars in my eyes. I had made it past the elephants and was hit by a buck! I don't know who got a bigger fright, me or the impala. We were both lucky not to be injured.'

The other rangers laughed uproariously at this story. Muzi joined in.

'What a day,' he said, shaking his head. 'I can only laugh!'

The rangers love to tease each other, but what fills me with happiness is that they all get along beautifully. There is no arguing or jealousy, they are respectful of each other and have great respect for Siya, as head ranger. I also know they all cover for each other when, for example, someone has had a couple too many beers! Over the years I have learned to close my eyes occasionally.

After some time as a tracker, learning from the guides, the next step up the ladder is to study for the professional exams of the Field Guides Association of Southern Africa, FGASA. And then – the final and usually the most difficult step – get a driving licence.

I was so proud to present them all with their FGASA certificates in March 2016. They were now professional game rangers, with a framed certificate for the wall, and a duty and responsibility to spread awareness about the plight of our endangered African wildlife. Our guests always leave Thula Thula saying they have gained great knowledge and under-standing about the bush – thanks to our excellent rangers. The rangers acquired their knowledge through long hours of observation, and learning from each other, not just from books. In fact, sometimes the rangers know better than the books! When Muzi was studying for FGASA, he was surprised to read that elephants never lie down, because they are so heavy and their organs would be damaged.

He told the assessor, 'I've seen an elephant lying down. I wondered if there was something wrong, but no, it was fine. So what do I say for this question?'

The assessor replied, 'You have to write what you find in the book, not what you see.'

This is why I am sometimes sceptical of so-called experts!

Many people dream of living and working in the bush with animals, and it truly is a great privilege, but it's not easy. It is a long day – the rangers are up at 5 a.m. and finish after the evening game drive. There is a lot of responsibility, for the guests, the animals and the vehicles. During Covid, they took on more responsibility for rhino monitoring and many other tasks they wouldn't ordinarily have to do, and I never heard one complaint, just, 'No problem, Madame' – or as they sometimes call me, Ma Fra.

The rangers – along with the staff at the Lodge and Tented Camp, of course – are the face of Thula Thula. They share their passion, knowledge and vision for nature and wildlife conservation with all our guests and volunteers. Because our rangers have been here so long, they have history with each other and with the animals. They recognize every elephant and know their name, history and character. It's the rangers who will notice if an animal is sick or behaving oddly. They are the first to see that a baby animal has been born, and to announce it to the rest of us. Each one of them is part of the story and the history of Thula Thula, working for the same goal, with the same passion and vision that Lawrence had, and they have helped me to carry on his legacy.

17

The Kindness of Strangers (and Friends)

Within days of our putting the rangers up for 'adoption', Portia came to me with a big smile: 'It's working!'

The adoptions started to take off. Our beautiful staff and animals had touched a chord with people all over the world and money was coming in.

'This lady Ellen Olson Brooks from Colorado has adopted all our dogs and our wildlife,' said Lynda.

'She is a great animal lover,' I said.

'And she's adopted every single staff member – even you Françoise!' The adoption of Madame caused much merriment amongst the office staff.

'Maybe one day you will be as popular as Baby Jo,' said Portia. 'Look how many people have adopted her!'

'Baby Sissi too,' said Lynda. 'Everyone loves her.'

'I can't be expected to compete with the baby animals,' I said. 'They are too cute!'

As for the dogs, Christiaan's big boy Bruce was super popular. People wanted to adopt him for real and take him home. Bruce was bred as a fighting dog – he looks like he's got some mastiff in him – but despite his size and his ferocious appearance, his temperament is placid. He's a lover not a fighter. Our mechanic found him abandoned in the street and took him in, and Christiaan adopted him to keep him company at Tented Camp.

He is a big, solid, gentle soul who gets on well with the rest of our pack and with humans – he reminds me of our gentle elephant giants in that regard. He's a favourite of visitors. Christiaan has got used to the fact that returning visitors will usually run to greet the dog with an enthusiastic, 'Ahh, Bruce, I missed you . . .' before he gets a hello himself. Christiaan was like a proud daddy when he heard that his dog was the most popular. I didn't tell the other dogs though; we don't want jealousy in the pack. And besides, every dog is special in their own way.

The rangers teased each other about their popularity or lack thereof. My lips are sealed on that subject – as far as I'm concerned, they are all very beautiful! What was very special was that, through the adoption programme, everyone at Thula Thula, from the rangers, to the elephants, to my little Gypsy, did what they could to help raise the funds to keep us afloat in this difficult time.

The rangers always tell me that working at Thula Thula is not 'a job', and it's true. I welcome their commitment and initiative in proposing new ideas for projects, or new solutions to a challenge. They take pride in their work and love their jobs because of this freedom to express themselves, and because they know that I have great trust in them and appreciate their ideas and input. No one succeeds on their own in life. And certainly not in the bush! We are all working in unity for the greater good, for the betterment of Thula

Thula and all our wildlife. I feel grateful for my Thula Thula family and the vision we share. We were humbled by the generosity and love, both from our guests and friends, and from strangers all around the world. People's open-hearted support kept us alive in the darkest times.

I was particularly touched by Ayan Mehra, a twelve-year-old student at the American School of Singapore, who is a very talented artist and a creative and determined fund-raiser. He made drawings of all our rhinos, of big Bruce and my little Gypsy, and many other animals. Ayan built a website to sell his art and used the money to adopt all our rhinos and most of our elephants. He also writes excellent blogs to raise awareness about wildlife conservation. Thinking about the bigger conservation picture, he is now raising money to contribute to our land expansion project.

Ayan's amazing work on behalf of Thula Thula caught the attention of the animal charity, Born Free, and in September 2021 he was made their first ever International Youth Ambassador. I am so proud that this young man's dedication and love for wildlife conservation was ignited when he read my first book, *An Elephant in My Kitchen*. He is such an inspiration for future generations. Ayan and his family were coming to visit us in April 2020, but Covid-19 made that impossible. His donations helped us so much in those challenging times – we hope to meet you one day Ayan!

The national alert levels were gradually lowered, and restrictions eased from August 2020. Europe opened up somewhat in the summer. At last, it felt like there might be some good news around the corner.

International travel remained all but impossible, but the easing of lockdown regulations did give us a tiny gap in which to operate. We were allowed to have guests under very particular circumstances – for educational purposes or business meetings. At last, something we could use! We put our heads together and dreamed up ways to have guests safely at Thula Thula and bring in an income.

The Thula Thula wildlife photography course – educational, of course – was one of the mainstays of the Lodge for a month or two. Students came for a five-night stay, for a mix of theory and practical photography skills from Kim, and a daily game drive where they had a chance to practise their newfound skills and knowledge and build a portfolio of wildlife pictures. Thula Thula is the perfect place for it, because the landscape is beautiful and photogenic, and our magnificent elephants come so close.

Soon regulations eased further and we were able to welcome South African guests – we assured them that fresh air, ventilation and social distancing were guaranteed. People were desperate to reconnect with nature after almost six months of lockdown in their homes. The guests couldn't be happier to be in the bush. The alcohol prohibition was lifted, which was a relief to many. There was a renewed sense of freedom all round.

It was wonderful to have guests with us again, and no one loved it more than the elephants – the visitors had come back to admire them at last. As the first game drive rumbled into view of the herd, Mabula, ever the attention-seeking entertainer, strolled into the road, where he posed and preened, showing off his yoga moves and trying out all his tricks for the tourists. Then he let out a blast of a trumpet. His mum, Frankie, looked on indulgently. She knew he was saying, 'Welcome back! Look at me! Look at me!'

18

What's Ours Is . . . Mined?

In August 2020, Vusi came into my office looking worried.

'Did you know there's a mining notice up at the entrance to the Lavoni place?' he asked. 'I was there this morning and I saw this.'

He handed me his phone with a photograph of the notice. I ran my eye over the legalese:

Notice is hereby given in terms of . . . . XYZ has applied for a Mining Permit . . . kindly submit any comments and concerns before . . .

My blood ran cold. This was not the first mining threat we had encountered. I knew that in terms of the law, anything below the ground belongs to the state. We might own the land, but the state owns – and can sell – the mining rights to someone else. If the land is not being utilized, they can exercise those rights and come and start digging. We had seen off a 2017 mining attempt on the reserve, on the grounds that we have endangered species living on that land. But this time, the area concerned wasn't yet fenced, so the land was regarded as unused. There were no wild animals on this land, so we couldn't use that argument. There was nothing to stop them coming to dig for coal in our backyard.

The area in question was the piece of the old Lavoni farm which had been a mix of bush and sugarcane, and which hadn't been incorporated into the reserve. The sugarcane company had had a lease until March 2020, and we had intended to fence the area as soon as the lease expired – another one of our plans scuppered by Covid. The mining company must have known that the sugar company's lease was up, and thought they would get in quick while it wasn't being used.

The struggle to survive the pandemic, and now this! A coal mine right on our doorstep would be devastating to the natural environment and to Thula Thula.

'They can't mine here, they will destroy the bush and terrify the animals with all that noise,' I said. 'Imagine the effect that the blasting and excavating and grinding would have on the elephants and rhinos. And the water and air pollution from coal mining is terrible. This could spell the end of tourism in the entire area.'

'It would be a disaster. We have to fight this,' said Vusi.

As soon as we'd finished talking, I called our excellent lawyer, Kirsten Youens. Specializing in environmental law, she has worked on many such cases in northern Zululand, stopping mining in environmentally sensitive areas. She's a brilliant lawyer and she speaks simply without a lot of legal-ese, which I appreciate. Kirsten promised to get in touch with the would-be mining company.

Meanwhile, I told Christiaan and Vusi, 'We need to get started on that fence straight away to make sure that area is recognized as properly part of Thula Thula, and therefore not suitable for mining.'

'OK, we'll get right on to it,' said Christiaan, who, together with Vusi, was in charge of the project. 'It's five kilometres of electric fencing though, that's a very big job . . .'

He was right. Fencing is physically demanding and time-consuming. For every post, a hole has to be dug in the hard, stony ground.

'I know, but it has to be done, starting now! Ask Jack to get involved.'

At the time, Jack was looking after the volunteer academy. The volunteer programme is an important part of what we do, but the fencing project was a priority, and we needed all hands on deck. The very next day Vusi, Christiaan and Jack were out surveying the fence-line to come up with a plan of action. Jack had a genius idea. 'The interns! Why don't we get the interns to help? They can stay at VA and put up fences.'

The travel restrictions in the pandemic had meant we were not getting our usual volunteers from around the world. By the European autumn, figures were not looking good in the rest of the world. Some countries seemed to be entering a second wave of infections. Any hopes I still had for the return of international travel were shattered. At the same time, many South Africans working in conservation and tourism had lost their jobs. We had welcomed some of them as interns, giving them food and lodging in exchange for work contributing to our expansion project, and doing other jobs, like monitoring our rhinos. Rather than staying at home waiting for tourists to come back, they got work experience, fresh air and adventure in the bush, a sense of purpose. And, of course, the opportunity to get to know our special herd of elephants! It was a win-win situation, and everyone was happy.

We put the word out, and soon had a team of enthusiastic interns from all over South Africa working on the fencing project. They ranged in age from eighteen to seventy-seven years old, and the more mature interns were sometimes the most energetic. I told Jack to go easy on a lady in her seventies, but she wouldn't hear of it. She insisted on being given the same demanding tasks as the younger interns, and she did those tasks with a great sense of purpose. We employed people from the local community too. We invited our wonderful, generous friends and supporters to sponsor a metre (or more) of electric fence. Everything came together. That fence was going to go up!

In October, we started fencing the new area of Lavoni, which was to be a six-month project. A bulldozer came in and cleared a five-metre swathe of bush along the border, where the fence-line would be. We dug 1,100 holes, each 1.5 metres deep, in the rocky soil, and planted a pole in each one. Then the wire went up, and then the electric fence, and the solar powered energizers to power the fence. We did all of this by hand. There are machines that can do it quicker, but we were watching the funds carefully as the future was still uncertain with the pandemic lasting longer than we thought.

It was early summer, and in our subtropical climate, it could be sweltering – into the thirties – and they were mostly working out in the blazing sun. I regularly drove out to chat to the people who were working so hard for conservation, to admire their progress and to express my appreciation for their dedication to this emergency project. On my visits, I liked to look at this lush new acreage and imagine the day when all of this land would be one, and we could welcome our elephant family into the new part of the reserve.

'Imagine how happy they will be to discover this river,' I said to Jack and Victor. The newly incorporated piece of Lavoni has a river which runs through beautiful thick riverine forest. It is humid and dense and quite different from anything that our elephant family is used to. Impenetrable by most animals, this virgin bush would be a feast for elephants, and by moving through the bush and breaking and eating it, they will open it up for the other animals to access its tasty treats.

Tall alien succulents called queen of the night have found a foothold in some areas. They are spiky and covered in beautiful white flowers that open up at night and attract bats, which spread their pollen. They produce delicious fruits and propagate easily.

'Don't worry about those,' said Christiaan. 'The elephants love them. They actually knock them down and then roll them on the ground to get rid of the spikes so that they can eat them. Fortunately, they don't eat tree euphorbia – they look similar to the queen of the night, but that milky liquid in them is bitter and toxic.'

'We'll have to keep an eye on what they are eating. We don't want to lose the cabbage trees,' said Vusi. Cabbage trees, or kiepersol, are unusually shaped sculptural trees with a round crown of grey-green leaves at the top of each branch. They're beautiful – and delicious to elephants. 'We might have to protect them. And the young marula trees too. Put a fence or stones around the base.'

'Yes. The sycamore fig trees should be safe though – look at the size of them,' said Christiaan, smacking his hand against the broad trunk.

These huge sturdy trees produce abundant fruits almost all year round, much to the delight of birds and bats and monkeys and, when they fall to the ground, a host of other buck, as well as smaller animals and insects.

There were already plans afoot to study the impact that the animals will have on this brand new land, which has not been home to elephants, or even smaller wildlife, for many years. A portion of the land along the river, and another in the open area, will be marked out. Each block will be carefully surveyed, to make a list of all the plant species and quantities. In a year, we will go back and recount, and we will be able to see the impact – positive or negative – of the animals' arrival.

'It'll be interesting to see how quickly those sugarcane fields return to the indigenous bush,' said Victor.

'We just need those elephants to walk about a bit, trampling the remains of the cane and feeding the soil,' said Christiaan.

'Two things you can always count on elephants for is trampling and creating dung!' I said.

Elephant dung is an important – and abundant – source of food for the shiny black dung beetle you often see heroically pushing a perfectly round ball many times bigger and heavier than himself along the reserve's dirt roads. These fascinating creatures – of which there are eight hundred species in South

Africa – have an important role in the ecosystem. The ball is made of dung which will be used as food, either for the beetles, or for the larvae which will hatch from the eggs the female lays inside it. The dung ball is buried in the ground where it decomposes, aerating the tightly packed soil and getting the nutritious elephant fertilizer below ground. By moving the dung in this way, the beetles also reduce the number of flies and other organisms around, and the larvae are a favourite snack of animals like mice and honey badgers.

The rangers moved on to another favourite subject – the new animals we will get as we expand. With expansion, you don't just get more land, you get different land, which means different plant and animal life, new insects and birds. Some we would have to introduce, others just arrive if the environment is right and they have the water, food and shelter they need.

'We might get animals like reedbuck, steenbok, bushbuck, warthogs, dassies . . .' said Christiaan.

'And if you bring in that kind of prey, the predators follow. We know there are leopards in the area,' said Victor.

'And eagles and buzzards,' said Jack. Where the other rangers tend to be all about the elephants and the rhinos, Jack is fascinated by the bugs and beetles, the snakes and lizards, the birds and the plants. He knows a lot about them and loves to share – hence his nickname, Dr Google. 'The area is extremely diverse in terms of plant life and with the new land we have all these different microhabitats, which means a high diversity of butterflies. Especially down here in the south, in the new part. Come summer, you see butterflies you don't see anywhere else. It's a real treat for any butterfly enthusiast.' He gestured broadly to indicate where the flocks of butterflies would soon be fluttering, revealing a trail of tattooed scarab beetles and other insects making their way up his forearm.

'If we get to 6 or 7,000 hectares, we might even get lion one day . . .' said Christiaan, his eyes shining in anticipation.

Predators – animals like lions, leopards and cheetahs which catch, kill and eat other animals – are important to keep nature in balance on the reserve. 'They take out the weak and sickly specimens,' said Christiaan, who is always very pragmatic – not a sentimental bunny-hugger like me! 'It leaves the stronger, fitter animals to breed, and the result is we have a much better bloodline.'

'And they help keep population numbers down,' said Vusi, casting an eye on a group of impala grazing on the open grassland on the hill across from where we stood chatting. 'These guys are doing a bit too well at Thula Thula. Too many more and they will put pressure on the bush.'

'With this good rain and the bush as lush as it is, the impalas can double their numbers in three years,' Christiaan nodded. 'It could be a problem. Now, if we had lions . . .'

The impala seemed to have caught wind of this discussion, moving quickly out of sight before anyone got any ideas.

Lion populations are in decline and – shockingly – the majority of the lions in South Africa live in captivity. An estimated 12,000 captive-bred lions are confined to cages so that humans can pet them, or hunt them, or otherwise make money from them. Through Four Paws, our partners in the rehabilitation centre, who are very active in trying to stop commercial breeding of lions in captivity, I have come to realize just how desperate the situation is for the so-called kings of the jungle. Animals are often kept in small enclosures, and they are not always well looked after. Their entire life cycle is one of suffering.

The cute and cuddly lion cubs are popular for petting zoos. But they don't stay cute and cuddly for long. After a few months of being petted by tourists or used as photo props, they are too big to be safely handled by the public. Adult females may be sold to breeders and bred intensively, producing multiple litters a year. Males who are less useful to breeders are shot by trophy hunters or killed and their skin, teeth, claws and bones harvested and sold, usually to Asian markets. Big cats that have been raised in captivity and handled by humans can never be rewilded. They will never be part of a pride, breed naturally and raise their own cubs, hunt in the bush as they should.

Most tourists are not aware of the problematic side of these big cat encounters. But the truth is, if you are able to interact with a wild animal, stroking or petting it, or having your photo taken, you are contributing to the problem. In any animal interaction, ask yourself, is this situation good for the animal?

If you love animals, instead of visiting a petting zoo or a place where wild animals are kept in small enclosures, rather come and have an authentic bush experience, in a place where animals roam free and wild and safe. Support game reserves that are looking after the natural habitat, managing animal populations responsibly, and caring for the animals and the people who work with them. Or visit one of our beautiful big national parks where your entrance fees will be put to good use in conservation. Instead of a quick selfie with a cub, you will smell the grass and the dust, hear the sound of the birds and crickets, experience the thrill of the game drive, and the excitement of spotting an animal, free and happy. It is an experience that will live in your memory for a lifetime.

It was because we want to see more lions out in the wild – as well as to manage the prey population – that we were considering perhaps getting lions, once our area exceeded the necessary 5,000 hectares. While we were discussing introducing lions or other predators to Thula Thula, I saw a fascinating documentary about the beneficial ripple effect that the reintroduction of grey wolves had on the Yellowstone National Park in the USA, on the other side of the world. These great predators were hated and feared by settlers, who shot or poisoned or trapped wolves to protect their livestock. The government encouraged their destruction, and by the late 1920s, they had been hunted to extinction. With the wolves gone, the deer and elk population had soared, resulting in overgrazing and erosion, and damage to the riverbanks.

In 1995, the first wolves were brought in from Canada and Montana. When the wolves came back, they controlled the elk population, taking out weaker and older animals and leaving a more resilient herd. With predators around, the elks' behaviour changed. They moved around more, so there was less intensive browsing. They avoided the rivers and gorges, where they were most vulnerable to the wolves, and those areas began to recover.

Other animals benefited. Young willow trees were particularly hard hit by the elks' population boom. When the willows recovered, the beavers had more food, and their population increased. Before the introduction of wolves, the beavers were down to one colony – now there are twelve beaver colonies. The beavers built new dams and ponds, which store water, and provide a healthy environment for fish and other water life. Regenerated areas attracted birds, mice, foxes and bears. And with more carrion around, the scavengers came back. So it was good for ravens, eagles, magpies and coyotes. The introduction of predators changed the whole ecosystem in ways that could never have been predicted, from the shape of the rivers to the height of the trees, to the kinds of animals that lived there.

I wondered what impact lions would have at Thula Thula. Having large cats around would certainly disturb the peace and tranquillity of our antelope herds who currently have little to worry about from predators. Lions could even take down a baby elephant, rhino or hippo. What worried me was that our wildlife have had no experience with lions but I knew that their natural survival instinct would kick in to protect themselves and their babies.

19

Big Plans for Big Cats

We have three leopards at Thula Thula, although they are seldom seen. Leopards are solitary rather than social animals. They come and go as they please, and they keep to themselves, lurking in dense riverine bush or rocky areas, coming out in the evening to hunt. If we are lucky, we catch sight of them on the trail cameras that are set up around the reserve as part of our anti-poaching strategy. It is always a thrill when I get a message on my phone at night and open it to see a picture of one of these magnificent beasts prowling through our bush. They are very strong and agile – a leopard can climb a tree carrying an antelope in its mouth. They are good jumpers and good swimmers, too.

These trail cameras offer some amusing glimpses of life in the bush. The giraffes sometimes find these mysterious metal blobs high up on a pole and come to investigate, giving us a curious view of an enormous eyeball fringed with magnificent lashes, or an inquisitive whiskery nose. We also spot nocturnal animals out and about in the day, and diurnal animals going for a stroll at night – they always look a bit sneaky, like kids skipping school to go and explore where they're not allowed.

Larry sent me a message one day:

Françoise, we have lost 3 cameras this past month. Broken by poachers.

Three? That is shocking.

Yes. They must be getting inside info. Someone is telling them where the cameras are set up.

That's not good Larry. Who could it be?

I will find out, I promise.

Two weeks later, Larry sent me another message.

Re the cameras. I've found the culprit.

There followed a photo – an extreme close up of an elephant's trunk reaching out to the camera. It was an elephant selfie – or should I say selephie? Either way, it was a very funny sight. Who needs TV, when you've got trail cameras?

Leopards are elusive and are usually only seen in brief appearances on the trail cams, but cheetahs tend to stay in a home range which the males mark out with urine. We didn't have cheetahs at Thula Thula, but I'd always admired these graceful, athletic animals – the fastest land animals – and harboured the hope of having them one day.

In 2019, Chantal Rischard and Stephan Illenberger had visited Thula Thula. When Christiaan mentioned to me that they had started the Ashia Cheetah Conservation in Paarl, near Cape Town, I was eager to meet them and hear about their work. I went to introduce myself and welcome them, and discovered that Chantal had read my book and they had come to Thula Thula to celebrate Stephan's birthday.

These were two remarkably passionate and visionary people. Their life plan was to work hard in Europe in the financial sector, make money, retire early, move to South

Africa and be involved in cheetah conservation. They had followed their dream to a tee! They now dedicate their lives to saving these vulnerable creatures from extinction by maintaining genetic diversity and increasing the expansion and resident range of the cheetah species. With the stars glinting above us in the *boma*, and a roaring fire warming our feet, Chantal, Stephan and I talked late into the night about life and wildlife, and conservation and cheetahs.

I knew that cheetahs were very vulnerable, but the situation was worse than I imagined – there are fewer than 7,000 cheetahs left on earth, down from an estimated 14,000 cheetahs in 1975. Like all wild animals, they are badly affected by habitat destruction, poaching and illegal wildlife trade, but Chantal explained that cheetahs have another problem as a species – they have very low genetic diversity.

'In South Africa, wild cheetahs mostly live in small and isolated populations in fenced wildlife reserves,' she said. 'Although the South African cheetah population is the only one growing in the past decade, gene flow is very limited and avoiding inbreeding is a real management challenge.'

To address this, the Endangered Wildlife Trust started something called the Cheetah Metapopulation Project, swapping cheetahs between the various reserves so that they breed with other cheetah populations from a different genetic line, and safeguard genetic diversity. Ashia works closely with the EWT in research, relocation and reintroduction of the cats. To find the right cats, a thorough background research into the history of each cat, as well as DNA testing are crucial. An extensive database of all the cheetahs ensures that the right cats get to the right reserves, thus avoiding inbreeding.

Ashia also takes in subadult mother-raised cheetahs with captive backgrounds and prepares them for life in the wild. The cats first spend a couple of months at Ashia's centre in Paarl where they improve their fitness on a special running field. They are only fed game meat and have the least possible exposure to humans. Once old enough to develop their innate hunting skills they are moved to one of Ashia's wilding sections, situated on two partner reserves in Northern Cape and Limpopo.

Chantal explained what happens: 'These pre-release or wilding sections are continuously restocked with prey species. Once a cheetah has shown to hunt successfully at least twice a week and is fully self-dependent, it is ready for release onto a private game reserve in South Africa or to become part of a reintroduction in national parks in Southern Africa like Malawi, Zambia or Mozambique to start new cheetah populations where the species has become extinct.'

I was impressed by their knowledge and passion for these beautiful cats, and I felt that we shared values. Like them, I believe that wild animals belong in the wild. Ashia had wilded and released more than twenty cheetahs, a wonderful accomplishment. I felt that we could work together.

As the embers of the *boma* fire died down, I said, 'Having cheetahs at Thula Thula was a dream that Lawrence and I shared. And having heard what you are doing, I would love to play a part in growing the numbers of beautiful species by giving them a home here.'

They suggested I speak to Vincent van der Merwe, the Cheetah Metapopulation coordinator of EWT's Carnivore Conservation Programme, about drawing up a predator management plan for Thula Thula. We did just that. The assessment of the reserve showed that we had the space, the variety of landscapes and sufficient prey animals to support cheetahs. In December 2019, we had finalized the predator management plan. The wildlife authorities require a minimum of 5,000 hectares of land in order to introduce cheetahs, and with our expansion plans, that shouldn't be an obstacle. I felt that cheetahs would be a better fit for Thula Thula than lions. They are not dangerous to humans, and they would give our wildlife a more gentle introduction into living with predators. I was concerned that lions could cause carnage.

We planned to introduce cheetah to Thula Thula in 2020. There was a beautiful female at Ashia who was ready to come to us as soon as the permits were in order. The cheetah – who we would later name Savannah – comes from an excellent bloodline, with high genetic diversity. Vincent gave us a rundown of her family tree. 'Savannah's gran was what's called a supermom. These are females who are fit and fertile, breed profusely and successfully raise their cubs to independence. We know from genetic testing that about ninety per cent of the metapopulation of four hundred and sixty individuals can be traced back to just three females, these supermums.'

Savannah's gran was captured in the Kalahari, and moved to Sanbona in the Western Cape, where she lived up to her supermum genetic lineage and produced twenty cubs. What's more, she successfully raised them in a tough environment that was home to lion – the predator responsible for about a third of cheetah deaths. Her daughter, Savannah's mother, was moved to a Garden Route Game Lodge. She also had the supermum genes, and she started breeding profusely, birthing fifteen offspring. One of whom was our beautiful cheetah.

Savannah's father was 'a strange cat', according to Vincent. He appeared at Mount Camdeboo on the Eastern Cape, and no one had any idea where he came from. 'He was spotted in the snow on the top of the mountain one winter. He must have escaped from one of the nearby reserves, Samara or Mountain Zebra National Park. He was a very wild cat, successful at avoiding lions, and a good hunter. He was unused to humans, and very clever – we couldn't catch him to collar him.'

They succeeded eventually, using a borrowed cage trap baited with meat – a capture method that rarely works for cheetahs – and he was collared and released. This wild chap proved very successful in breeding with the other cheetahs on the reserve. This was good in terms of numbers, but also problematic – there would soon be too many cubs with his genes, and there was a danger of inbreeding.

'He was moved to the Garden Route, and immediately got going there, fathering cubs,' said Vincent. 'Coming from the Eastern Cape, he had different genetics from the cheetah there, which is great. One of those cubs was Savannah.'

Hearing about this family tree, we knew Savannah was a wonderful animal with a very strong genetic lineage. At twenty-four months, she was at the right age for producing offspring, with high chances of also being a fertile, successful breeder. Of course, we needed a mate – or two – for her. We found two males from a different reserve and a different gene pool – an important consideration in a tiny population. Our plan was that Savannah and the two males would produce cubs which could be sent to other reserves, to help increase the genetic diversity of the population of these magnificent animals. All we needed were the permits to introduce the cheetahs, and the permits to move them across provincial borders to KwaZulu-Natal.

20

The Clever Clean-Up Team

Scavengers are just as important as predators. Hyenas and vultures don't get the good press that leopards and lions get, but they perform a vital function, cleaning up the corpses of dead animals. A cheetah or leopard might eat one hind quarter of a buck that they've killed, and – were it not for the hard working scavengers – the rest of the animal would lie rotting on the ground, a perfect breeding ground for diseases and parasites.

Out by the river crossing behind the Tented Camp, big grey birds perch in the trees, or set out in search of food, or even wheel in the sky, circling a carcass that will soon become their dinner. These are the white-backed vultures (*Gyps africanus*), and we have one of the biggest breeding colonies of these endangered birds in the country. Although it's the most common vulture in Africa, and one of the most widespread, the species is threatened by accidental and deliberate poisoning, trade, and the decline of habitat and food.

As scavengers, vultures are reliant on carnivores to kill larger animals and break up bones and hides, so when carnivores' numbers fall, theirs do too. As a result, these birds are on the 'critically endangered' list, and there are fears that they will go locally extinct by as early as 2034.

Christiaan found a vulture standing in the middle of the road to Tented Camp, head hanging down. It didn't fly away when he approached – a clear sign that it wasn't well. He picked it up and brought it home and tried unsuccessfully to tempt it with water and food. It really was a sorry looking bird.

'It looks like it could be poisoning,' he said. 'Farmers sometimes put poison inside dead goats to kill scavengers like hyenas and vultures because they think they kill their livestock.'

I shook my head at the stupidity and cruelty of our species.

'Sometimes animals are poisoned accidentally,' he continued. 'Hunters use lead bullets instead of the more expensive copper ones, and if they wound an animal and it later dies, scavengers can ingest the lead and die.'

Our rescued vulture really wasn't looking good, so we contacted the Vulture Group, who have expertise in taking care of these birds. They took the vulture to their centre and ran blood tests and came back to us with a diagnosis – avian malaria.

'It's very rare, like a one in a million thing,' Christiaan reported back. 'Anyway, the guy's in hospital and on the mend. He'll be fine.'

That was one lucky vulture!

Not everyone understands or appreciates the vital service scavengers perform, which is why they are hunted and poisoned to the point of being endangered. If only people understood how important these animals are, maybe they would be more inclined to look after them. That's where education comes in.

Victor is passionate about working with local communities to talk to them about wildlife and the environment. He has a calm, non-judgemental manner, and listens rather than lectures, so people feel free to share their beliefs – like the myth that if you dry out a vulture's brain, and sleep with it under your pillow, you can see the future, and even dream of the winning lottery numbers. My common sense tells me that if the vulture brains really helped you see the lottery numbers, the guy selling the vulture brains would be a very rich man! He'd be driving his Ferrari around Buchanana, not flogging dried out bird brains to the villagers. But common sense is not that common, is it?

Victor asks the kids, 'Who knows anyone in the village who has won the lottery?'

No hands go up.

'Ah, so do you think that it works to put the vulture brain under your pillow?'

There is some doubtful headshaking.

He shares some of the amazing true stories about vultures – that their sharp eyes can spot a dead animal from three kilometres up in the sky, that they sometimes eat so much so quickly that they can hardly take off, that they urinate on their own legs to kill off parasites. And most of all, he spreads the message that the environment belongs to animals as much as to us.

Hyenas share the vultures' undeservedly bad reputation

– people regard them as dirty and ugly and a threat to live-stock – and in fact, the two species are often found together, feeding side by side unless the hyenas feel that there's too much competition, in which case they'll chase the big birds away. Together, they are the 'clean-up team', and essential to the ecosystem.

At about the time that the dogs and I settle down for the evening, the spotted hyenas are waking up and going about their business. Because they are nocturnal, we only see them occasionally, on an evening game drive, or in the early morning. But, goodness, do we hear them! Hyenas are the most vocal animals with a big range of different sounds. They have a cackling 'laugh', which is not evidence of a sense of humour, but actually a sign of stress or nervousness, or submission to a dominant member of the clan. When they greet each other they make all sorts of groans and squeals, and they also grunt and growl. Every vocalization means something. The most common sound is a 'whoop', a call that travels for up to three miles, and is their means of long range communication – to call wandering cubs, or to bring the clan together, or to let others know that this territory is taken. It always fascinates me to think that the same loud whoop I hear from my sitting room is being heard at the Lodge and the Tented Camp. We are all in their territory, after all!

Hyenas are fascinating creatures, highly intelligent, sociable, and full of surprises. They look somewhat similar to dogs but are more closely related to cats. They are known as scav-engers, but they are actually very good hunters, too – let's just say they are not fussy! With their huge, strong jaws they have plenty of options – they've even been known to take cooking pots from campers' fires, or drag off crates of food. Their front legs are longer than their back legs, which accounts for their sloping profile. There's a longstanding belief that hyenas are able to change gender, but that's not true – people were probably confused by the fact that the males and females look so very similar.

They live in clans of up to one hundred members, highly organized and hierarchical, under the leadership of a matriarch. The females are larger, more aggressive, and more highly ranked than the males. An individual's rank determines when they hunt and how they feed. Dominance is passed down by the alpha female, through her female line, to her cubs.

Life is tough for hyenas, and their struggles start early. There are usually two cubs in a litter, and they compete for dominance (and their mother's milk) from the word go. After a few weeks in a den with mum, the hyena cubs move into a communal 'kindergarten' and there they learn to fight for position in their community. Their only predators are lions and – I'm sad to say – humans. Hyenas are an underappreciated species, under threat when living outside of reserves and protected areas.

In their organization and family structures they have much in common with elephants – no wonder that these creatures are favourites amongst our game rangers, who are always pleased to see them. Christiaan maintains that hyenas are generally quite harmless. 'They are inquisitive. If you sit down, so that you are at their level and your posture is submissive, they will come quite close.'

This, I have not tried!

Similarly, he says, if you lie down, a cheetah will likely come and sniff you to investigate.

Another suggestion I've yet to put into practice.

The rangers have all sorts of theories about what to do if you meet an animal in the wild, but everyone agrees with the advice: 'DON'T RUN!' It is of course anyone's natural instinct – to get away! But the movement triggers a chase response – you become prey. 'If you ever come across a lion, you just stand still. It will move away eventually,' they'll say casually.

Again, I hope not to have to put this to the test.

21

Is Something the Matter with Our Matriarch?

Like most of the world, I was pleased to see 2020 drawing to a close. It had been a long, hard year, culminating in a second wave of Covid infections in South Africa, with a frightening increase in hospitalization and deaths. Renewed restrictions were inevitable. What a year it was! In a couple of weeks it would be 2021. I looked forward to starting afresh, and to a better year.

I was in my office with the fan on full blast, taking shelter from a sweltering Zululand summer's day, and flicking through the photographs that the photography students had taken on their up-close encounters with the animals. As always, there were dozens of videos and photographs of the magnificent Thula Thula elephants. I never tire of looking at them. There's always something new to see.

I clicked on a video, and there was Frankie, approaching the vehicle where the student was filming. Now, Frankie had a strong, majestic bearing. She was the matriarch, after all, and it showed in her posture, her attitude, the way she carried herself, the way she moved. This was not the Frankie I saw on the video. She approached the vehicle in a way I'd never seen before. She seemed weak and needy, even desperate. Her eyes were sad. This was not our Frankie, the bold and proud and a little bit intimidating Frankie that I knew.

I asked Siya to arrange for all the rangers to come to my office for a meeting the next morning.

'Come and look at this,' I said, pushing play on the video. 'Look at Frankie.'

They leaned in to watch.

'See the way she leans against the vehicle and rests her trunk on the bonnet,' I said.

'It looks like there's something wrong with her,' said Khaya.

Andrew agreed, 'She looks weak.'

'She doesn't look right,' said Muzi.

'Perhaps she is sick,' Siya said.

They all saw the difference in her behaviour and attitude, even character. They all looked worried. Their responses confirmed my suspicion that something was wrong.

Elephants do get sick – they even get some of the same illnesses we do, like pneumonia, tuberculosis, constipation, colic, and even colds. Intelligent creatures that they are, they will seek out specific plants, indigenous grasses or herbs, or the bark of particular trees to help their digestive complaints, and they even cover small wounds with mud to keep insects and worms away.

I once saw Nandi treat baby Themba with a curative plant. Now, Nandi is a real sweetheart. In every family there's an auntie who always worries about everyone, fusses over them, and makes sure they're happy. That's our Nandi. She watched Themba make a poo, and then walked over and kicked the poo and smelled it with her trunk. A couple of other females came over, and they had a meeting, nodding and rumbling. Some manner of agreement seemed to be reached, and Nandi walked off into a thicket to pick a branch from a specific bush. She came back and handed it to baby Themba, just like an auntie giving a baby a kiddie's aspirin.

But of course there is only so much a wild animal can do to help itself. The way Frankie approached the vehicle, it was almost as if she were seeking help. It reminded me of how Frankie and Marula had asked for help when newborn Vusi was caught in a snare. Could this be the case with Frankie? Might she be asking for our help – for herself this time? Concerned for Frankie's well-being, I consulted our vet, Dr Trever Viljoen. I sent him the video and explained my concerns.

'What could it be?' I asked. 'Do you have any idea?'

'It's hard to tell,' he said. 'Although I can see she doesn't look good, I'm afraid it could be serious. Perhaps her liver or kidneys. It could be an infection, something she ate, or it could be something worse, like a tumour. It's very hard to say without full blood tests.'

It's not easy to determine what is wrong with an elephant. You can't just examine her, or ask her where it hurts, or send her off for X-rays. To do an examination and take samples for testing, it's necessary to anaesthetize her. This requires a helicopter, first to chase the rest of the herd away, and then to dart from the air. You don't undertake such an operation lightly – it costs a huge amount of money, and it's also risky. There is always the worry that the animal will fall badly, or respond poorly to the anaesthetic, or injure themselves when they wake up woozy. This puts the vet and the capture team in danger too – you don't want to be around four tons of woozy elephant.

'If you can collect some dung, we can run some tests,' said Trever. 'That might tell us something.'

Collecting dung samples is a lot easier than collecting blood samples, but it's not trivial. You have to follow the animal and wait for it to poo. Luckily elephants eat a lot and poo multiple times a day. I asked the rangers to be on the lookout, and to bring dung to be analysed. Muzi was the first to come in with a sample, which we packed up in a sterile jar and sent to the lab, as per Trever's instructions.

'The results are back but they aren't very helpful, I'm afraid,' Trever said, when he called a few days later. 'There's nothing obviously wrong.'

'What do we do next?' I asked.

'Let's give it a bit more time, see if she improves. Ask the rangers to observe her closely and be on the lookout for any changes in her behaviour.'

'Will do. We'll keep an eye on her.'

We all hoped that Frankie would rally. But in my heart, I was sick with worry. The matriarch is crucial to the success and well-being of the elephant herd, the one who keeps unity and peace in the family. She takes decisions on behalf of the group, deciding when and where the family feeds, sleeps and travels. The entire herd follows her lead, with total respect and discipline.

Nana, the first matriarch of the Thula Thula herd, was

gentle, yet firm. She could calm a group of rowdy teenagers simply by laying her trunk gently across the back of the ringleader. A nudge here, or a stern look there were usually all that were required to maintain peace and order. No need for fuss or conflict. When Frankie took over, she had her own leadership style. She was proud and strong, and was feistier than Nana. Frankie took no nonsense. She introduced her own rituals. When the weather was cold and windy, she would lead the elephants down south to Lavoni, where the deep valleys offered shelter. In hot weather they headed to their swimming pool, Mkhulu Dam in the north, for a cooling splash. Frankie even had a boyfriend, Gobisa. What a beautiful loving couple they were, always together, kissing and touching each other with their trunks. We hoped we would one day be welcoming their baby elephants into the family.

Life was good for our herd under Frankie, and the elephants were calm, peaceful and healthy. She was a good leader. With any luck, she would be our matriarch for some decades yet.

Every day, I asked the same question: 'How is Frankie?'

Every day, some version of the same answer: 'She doesn't look like her old self.'

Then one day, Muzi came back from an afternoon game drive and reported something very concerning: 'I saw Frankie. There seems to be a lump under her stomach.'

Photographs confirmed it – there was a noticeable swelling. I sent pictures to Trever who said it was an oedema, a swelling caused by excess fluid, but what it meant and why it was there, he didn't know. We decided to watch her carefully. Darting any animal is a risky operation. Darting the matriarch is extremely traumatic for the entire herd, and potentially very dangerous, given their unpredictable reactions. Over the next few weeks, Kim took countless photographs which we examined closely. The oedema appeared to be growing. Even more concerning, her skull seemed bony, the top strangely sunken.

At the end of November, I was at the Mkhulu Dam with a French TV production crew who came to film a programme on Thula Thula. It was a beautiful bushveld evening, the setting sun glinting off the water, bathing the herd in a golden light. Frankie was at the edge of the dam, her son and daughter Brendon and Marula close by, her devoted Gobisa at her side. I watched as he tenderly put his trunk over her, as a human might put an arm around a loved one's shoulders. It seemed a gesture of protection or compassion, as if he was telling her, 'I am here for you Frankie'. It seemed her close family knew she was suffering and was surrounding her with their love and protection. It was so beautiful and touching to see but it was also sad. I feared for Frankie.

22

A Christmas Visit with a Mission

On Christmas Day 2020, instead of reindeer, we were visited by our magnificent herd of elephants.

Our herd has a remarkable sixth sense and has been known to appear at the house at significant times. When Lawrence died, the elephants arrived, almost as if they knew their protector was no longer, and they wanted to see me and acknowledge his passing. For three years in a row, on the anniversary of that day, the elephants came to the house at exactly the same time, to pay their respects.

I can't give you proof or a scientific explanation, but I have witnessed for myself this amazing phenomena. I don't know how it happens, but I do know that elephants have deep intuition, an almost spiritual awareness, and that they have strong attachments to the humans in their lives. When they come to the house like that, they always have a message for us.

So when the whole herd arrived on 25 December, it wasn't so very strange to imagine that they had come to wish us a happy Christmas. I walked down to the fence to greet them and I realized straight away that something was amiss – Nana was leading from the front, and Frankie was nowhere to be seen. The elephants had come to show us that Frankie was missing.

It's always a worrying sign when an animal leaves the herd – it shows that she knows something is wrong, or even that it's time to die. And for the matriarch to be absent is unheard of. I immediately called Andrew, who knows a lot about elephant behaviour, and who also seems to understand them intuitively, on a deeper level – I often think he communicates better with elephants than with humans. I hoped he would be able to help me figure out what was going on with Frankie.

'Andrew, I am very worried. The herd is here but there's no Frankie. Where could she be? And why would she leave the herd?'

'Sometimes they isolate themselves if they are not well. She could be sick. Or injured. Or . . .'

There was a pause, as neither of us mentioned what was in our minds – that she might be dead.

He continued, 'Maybe if Frankie is weak, she doesn't want to slow down the herd and possibly compromise their safety.'

I had another idea: 'Or maybe she knows that being on her own makes it easier for us to check on her, to take care of her?'

'We need to find her,' he said.

The search was on to find Frankie, who we now believed to be in real danger. Time was of the essence. On Boxing Day, the rangers fanned out, searching her favourite places, the water holes she frequented. They sought her everywhere. It was a huge relief when our head ranger, Siya, sent me a message: *I've found Frankie. She's alone and she's alive.*

Frankie had left the herd and travelled south, always her favourite area of the reserve. We climbed into the game spotting vehicles and headed out, following Siya's directions. She was exactly where he'd seen her hours before, in the middle of a road.

The fact that she hadn't moved was not a good sign.

On the other hand, the fact that Frankie was alone and in an open space meant that it would be easier and safer to dart her – you can't dart an animal in thick bush, and the presence of the herd makes any operation much more perilous. I called Trevor and we decided it was time to tranquillize Frankie, do more extensive tests and get some answers. Dr Viljoen arranged for a helicopter and all the necessary medical supplies. The date was set for the very next day, 27 December.

The question of whether to intervene when a wild animal is sick or injured is always a difficult one. We usually let nature take its course, but I make an exception if the problem is caused by humans – as a result of poaching, for example. In that case, I believe it's our duty to help the animal. In Frankie's case, I took the decision to interfere with nature even though this was not a man-made injury. This was Frankie. She was our matriarch, an important member of the Thula Thula family. I could not sit by and watch her fade away before my eyes. I had to at least try to save her.

Our whole team was there for the darting – all the game rangers, Christiaan and Lynda, Kim filming and photographing to record the operation for our archives. Darting a huge creature like her is risky, and I wanted the whole team there in case of a problem. It's also an educational opportunity for everyone. With all our rangers and guards in vehicles on the ground, and Dr Viljoen in a helicopter, we brought this great beast safely down. We all rushed towards her.

It was heartbreaking to see Frankie lying there in the grass. She was massive, of course, but even so she seemed thin and vulnerable, her great rib cage rising and falling as Trever pressed a large stethoscope to her chest and then her abdomen. Her huge, inquisitive trunk sprawled lifelessly on the ground next to her, with a little stick propping it open so she could breathe.

I reached out and stroked her broad shoulder. It was the first time in all these years we had known each other that I'd laid a hand on Frankie. Her skin was deeply wrinkled, rough as tree bark and bristly to the touch. My hand came away lightly covered in sand and dust. As I leaned over her, touching her, I was also praying and whispering: 'Don't leave us, Frankie. You can get better; you must get better. Please don't leave me. Don't give up.' Who knows what Trever must have thought of this mad woman talking softly to this four-ton creature as if to a sick child.

Trever worked quickly, starting with a physical examination. He opened her eye and pulled down the lid to show me.

'Look,' he said. 'Her eyes are inflamed. Can you see? And the inside of the eyelid is very pale, it should be a darker pink.'

He pulled her lip to expose the inside of her mouth.

'The gums too, look . . .'

They were pale and greyish against her huge yellow molars. I knew enough to recognize that this wasn't a good sign.

He took big vials of blood from the veins in her ears and then took out a couple of huge syringes.

'What are you giving her?' I asked.

'This is a general antibiotic,' he said, plunging the needle into her massive shoulder. 'If it's a bacterial infection of some sort, the antibiotic should help.'

He picked up another giant needle. 'These are vitamins and boosters. Until we have more information, we can just hope that one of the medications has an effect.' He sunk the needle into her thick grey hide.

While Trever prepared to inject Frankie with an antidote to reverse the effects of the anaesthetic, the rest of the team decamped to the top of a nearby hill a few hundred metres away. You don't want to be too close to a drowsy elephant staggering around after a dose of tranquillizers!

Frankie lay there for a few minutes and, just as I was beginning to worry that something had gone wrong, she got to her feet, standing strong and proud, head high. She looked good, majestic, the old Frankie. We all cheered and clapped, feeling relieved and hopeful, as Frankie walked slowly away into the bush. It had been a very long day and I was exhausted by the emotion – my fear for her health, and the stress of any animal tranquillizing operation.

I thought of my long history with Frankie. She was so intimately involved in our story – mine and Lawrence's and that of the herd. She was only forty-six, just middle aged in elephant terms – she should have another twenty years with us.

I hoped with all my heart that whatever was wrong with Frankie was treatable and that she would recover to lead the Thula Thula elephants for many more years. There was nothing to do but wait for the results of the tests.

23

Oranges for The Queen

'Still no oranges?'

'I'm afraid not,' said Clément, unpacking the shopping onto the kitchen table. 'They are out of season. I got apples though.'

'The Pink Lady variety?'

We both smiled. Frankie's preference for high quality fruit and vegetables had become a joke at Thula Thula in the days that we'd been feeding her. The tests showed that Frankie had a malfunction of the liver. Dr Viljoen wouldn't give a firm prognosis – he muttered something about '50/50' – but I was feeling optimistic. We at least knew what was wrong now. We could start moving forward, and try to build up her strength. With my ever-positive outlook, I am always prepared to put energy and effort into looking for solutions and getting a good outcome.

Frankie was not eating well, and was looking thin and weak. The rangers fed her lucerne – which elephants adore – but she just played with her food, tossing the delicious sweet grass onto her head and letting it fall to the ground. They had better luck with horse pellets, which are like a nice big bowl of tasty cocktail nuts to an elephant, but even so, her appetite wasn't what it should be. I took it on myself to feed Frankie back to health. After all, look at how chubby all my skinny rescue dogs were! I would build her up and help her get her strength back. And make sure she took the medicines and special liver tonic Trever had prescribed.

I prepared two meals a day for Frankie, a task I did with love, with some chopping assistance from my sous chef, Clément. In addition to the horse pellets, I mixed up a big bowl of fruit and vegetables, hoping to give her a good variety of nutrients, and experimenting with various options to see what Madame enjoyed. Elephants love bread, so I added a few slices as a little treat until the vet told us not to. A typical meal would look something like this:

Six apples, halved
The leaves of a whole cabbage
A couple of oranges, if we could find them
Other bits and pieces – cucumber, tomato, a few
  grapes, whatever we had
A good glug of the vet's liver tonic
A teaspoon or two of cannabis oil, because after all,
  who knows?
All mixed in with few scoops of horse pellets

This delicious mixture was handed over to Andrew and Siya, along with stern reminders to bring my big salad bowl back! We wanted to minimize the number of different people Frankie came into contact with in order to keep her as calm and stress-free as possible. Andrew was our elephant expert, and he and Frankie had a beautiful bond. Siya was our head ranger, I trusted him completely and he was brilliant with animals. One of them fed her in the morning and the evening.

This was not a trivial task, because in order to feed an elephant you have to find the elephant! How hard can that be? you ask. Well, very! One of the puzzling things about elephants is that despite their enormous size, they are notoriously difficult to locate. They are surprisingly quiet, even as they move through dense bush. We discovered this when the elephants first arrived at Thula in August 1999. They hadn't been there twenty-four hours before they escaped. Lawrence and I drove the dirt roads of Zululand for a whole week before we spotted them! Even in her weak state, Frankie could move quite long distances. With one of the rangers driving, Kim taking photographs and videos to monitor and record her progress, and Siya or Andrew in charge of the food, finding and feeding Frankie might take seven hours a day.

My, she could be quite the princess, our Frankie.

When the lovingly prepared mixture was deposited in a bucket or in a pile, she sauntered up and nosed about in it with her trunk, looking for treats. If she found a nice bit of orange, or a slice of Golden Delicious apple, she'd pop that into her mouth and crush it with her great molars. Then she'd push the food around some more with her trunk, to see what else was on offer, picking out only the choicest morsels. She made it clear that she did not appreciate the cabbage which I'd included to boost her iron intake, delicately picking out the cabbage leaves and tossing them to one side. One day, I happened to have half an organic gourmet baby cabbage from

Woolworths in the fridge, which I added in, and she ate that. What a madam!

Just a day or two after the feeding started, it seemed as if we were seeing results.

'She's looking a bit better, don't you think?' I said, looking at the video Kim had taken that morning.

Sure enough there was Frankie looking a bit perkier, more like her old self. We passed it round – Vusi, Christiaan, we all thought we saw a bit of an improvement. Andrew seemed a little less convinced and Kim held her tongue.

The next day it rained. We usually keep the windscreens of the game vehicles folded down onto the bonnet, and rain water had collected on it. Frankie was delighted to find fresh water – elephants will drink what water they can find, but they love nice clean water if they can get it. If they get so much as a whiff of fresh water, they'll happily dig up a pipe to get at it.

One memorable New Year's Eve, while the champagne was on ice for the guests, our herd decided to celebrate with delicious fresh water. I was helping Zandile, the chef of Tented Camp, with the potato and onion rösti that would be topped with smoked salmon with a chive and lemon chantilly, when the guests arrived back from the evening game drive and found their showers dry. I asked Vusi to investigate and he soon found the culprits. He called us to come and see. The elephants were gathered around a water pipe just outside the fence, taking it in turns to enjoy the sweet water and the growing mud bath that was seeping towards us. The guests who were seated at the outside bar watched the muddy elephants drinking from the pipe. They were charmed, except for one lady who'd already donned her long dress and heels (not exactly bush attire, even on New Year's Eve), now rather muddy. It is no fun trying to cook a special meal and keep the kitchen clean and tidy when the elephants have hijacked your water supply. But we managed, and the midnight arrival of champagne helped and Zandile and I started the year in a very happy mood.

Frankie slurped up the water pooled on the glass as if it was a delicious glass of sparkling Perrier, and from that day, we brought her fresh water with her food, splashing it onto the windscreen so she could suck it up with her trunk.

Over the days and weeks, Frankie's personality changed. As long as I've known Frankie, we have all kept our distance from her, and treated her with caution and respect. She wasn't aggressive, exactly, just a little, how would you say . . . intimidating.

Every day I asked Andrew, 'What do you think? Does Frankie look any better?'

'Some days,' he said, avoiding my eyes. 'Maybe a bit . . .'

'Do you think the medicine is working?'

'I don't know Françoise . . .'

'I won't give up hope. We must be positive!' I said.

I went out with Andrew to see her for myself. I held the bowl of food as we drove out to find her. She was browsing peacefully just metres from the road. Andrew took the big bowl of her special treats from me and got out, calling her softly, 'Hey Frankie, come on girl . . .' She walked towards him at a slow pace, almost dragging her feet. I hardly recognized her. Her manner, her comportment, her walk – all were unfamiliar. But the most upsetting change was in her sunken skull and her dull eyes. Frankie ate her special lunch, tossing the despised cabbage leaves contemptuously from the pile, and approached the vehicle. She walked slowly past us and stopped. We stayed there, Andrew, Frankie and me. Having isolated herself from her own elephant family, she seemed to want our company. It felt as if she was trying to tell us something during this long visit on the road.

After some time, Andrew started the vehicle and moved slowly away. Frankie stood there on the side of the road, watching us leave. My heart broke to see her there all alone, looking so old and frail. As devastated as I felt, I still refused to accept reality. I told myself that Frankie would recover, that she would return to the herd as the mighty matriarch. That order would be restored, and life would go back to normal.

Even my determinedly positive approach was shaken by her deterioration in the first few days of January. Siya and Andrew reported that she was starting to slow down. Her skull seemed increasingly sunken and the oedema seemed to be growing. Kim's videos and photographs confirmed these facts. We pored over the daily pictures and emailed them to the vet. Trever wasn't happy – the growth of the oedema was a sign that the liver wasn't working as it should and water was accumulating, being drawn down by gravity to the lowest point, beneath her belly. I couldn't admit even to myself what I was seeing. I still held onto the belief that she would get better. The thought that she might not was unbearable. But

I started giving her bread again. What harm can it do now? It was time to let her have her favourites.

And then, she was on the move. From her favourite places in the south, she started walking towards Tented Camp. What a good sign! I was filled with fresh hope that Frankie might be turning a corner. The next day, on 8 January, Andrew found her at Croc Pools. He got out of the vehicle, walked towards her and put the food down. She looked straight at him but didn't move to the bucket of fruit and treats.

'Frankie,' he called, shaking the horse pellets to make a rattling noise. 'Come Frankie.'

She moved away a few steps. He moved too, bringing the bucket closer.

'Come on Frankie. Come on girl,' he said, reaching into the bucket and pulling out an apple.

Frankie gave her old friend one last, long look, and then turned and walked slowly away without eating any of the lovingly prepared lunch. She walked unsteadily, swaying as she crossed the dam wall. Andrew watched, apple in hand. Frankie stopped on the shady bank on the other side. It was a good spot. She had food and water nearby. She could stay there and rest in the cool tambotie forest. We would know where to find her the next day, when it was time for breakfast. Perhaps she'd have more appetite then.

But we couldn't find her the next day.

For more than a week we looked on foot, in vehicles, with drones. Everyone was alerted – the rangers, security, the Anti-Poaching Unit. We were all looking for Frankie. We couldn't find her. She was so weak, she couldn't have moved far from the place she had last been spotted. How could we not see her?

I had to go to Durban for a few days for medical tests. I intended to return on Sunday, but on Saturday, I had the strangest unsettled feeling. I needed to get back to Thula Thula. I got in the car and drove the two hours home.

Kim was sitting by the pool when I arrived. I joined her to catch up on what had been going on. We looked into the reserve across the fence, our eyes gazing out across the bush.

'Where is she . . . ?' I asked, into the air.

Kim knew, of course, who I was talking about.

'Where are you Frankie?'

Kim shook her head, 'Everyone's looking.'

'Maybe she'll come to us. She's sick, she might come close to the house, knowing we can take better care of her.'

'Maybe. Maybe tomorrow.'

'We can't give up hope.'

Lynda came out of the office, walking briskly towards us. As she got close, I saw that her face was grim, her eyes swollen from crying.

'The APU has just called,' she said. 'They've found Frankie.'

The unspoken question hung in the air.

'Frankie is dead,' Lynda said quietly.

There was a shocked silence as we struggled to take in this terrible news.

'*Ce n'est pas possible*,' I said in agitation. 'It can't be true.'

But of course, I knew in my heart that it was.

'Where is she?' I asked.

'The south.'

'Let's go,' I said. I wanted to see her. I needed to see her, to really believe that it was over, that Frankie was dead. It was unimaginable that she was gone. I called Andrew, my voice breaking. 'Please come now, I have to see Frankie. And we have to go quickly. It's getting dark.'

He didn't argue. Minutes later he arrived. I climbed in next to him, tears in my eyes. Kim and Lynda jumped in the back. We drove in silence as the sun dropped towards the horizon. We met Mandla, alone, close to the house. A little further down, Nana, also alone. Then Mabula, also on his own. As we passed each of them, they stopped grazing to stare at us as we passed. It was unusual for the elephants to be dotted about like that, strange for them to look up at us and follow us with their eyes. I couldn't help but wonder, did they know about Frankie's death? Did they know we were on our way to her? The tears started to fall as we passed the dispersed herd. What would happen to them without their beloved leader? I cried quietly the whole way there.

The APU team was on the phone, giving directions. Andrew pulled up at a place of thick, almost impenetrable bush. The APU guys were waiting to show us the way down to the little dam where they had found Frankie. We got out of the vehicle and walked through savage thorn trees and long grass dotted with ticks. There was no road or path. Only the smell of her decomposing body gave away her location and that's what brought the APU men to her, days after her death.

My tears blinded me. Andrew pushed the thorn branches out of my way, and Lynda and Kim had to help me walk. Thorns snagged and caught on my shirt and scarf and scratched my arms, but I was numb with grief, feeling nothing. The smell became stronger the closer we got, and soon it was overwhelming. I pulled my scarf over my nose. I heard the buzzing of flies.

Andrew stopped.

There she was, our matriarch, lying on her side as if she were sleeping peacefully. I collapsed, overcome by the reality of it, the finality. She was gone. Frankie had chosen her forever place, a place so remote that we had never set foot there, or known of the little dam she lay next to. There was no path to her resting place. She had left us with the dignity, pride and humility of a true leader.

We left her body there, taking only her tusks to save them from the poachers. Later, Christiaan went back to fetch her skull. Poachers had already hacked at it to take pieces of bone for *muthi*, or traditional medicine.

We knew that the herd would find her and come and say their own farewell. In death, she would continue to be part of the great cycle of life, food for other animals great and small, for hyenas and vultures, insects and microbes. She would return, eventually, to the soil.

The drive back to the house was the saddest journey I had ever experienced. We drove in silence, as dusk settled quickly upon us, each with our own thoughts, our own memories of Frankie. We had so much history together, Frankie and I, and I thought we would have so much more. How could she leave me? Fresh tears flowed when I thought of her beloved Gobisa, and of the baby I had so hoped they would have together. My dream was lost.

What would happen to our elephant herd now?

24

Savannah

Frankie's death was a devastating blow to all of us at Thula Thula. The immense grief we felt at the loss of our matriarch seemed to suck all the joy out of daily life. While mourning Frankie, I was also plagued with worry about the future of the herd. How would they cope? Who would guide them? I wondered at their own emotions – I had no doubt that the elephants were grieving, as we were.

Just days after Frankie's death, we would see the fulfilment of a long-held dream. On 19 January 2021, the female cheetah was coming. The introduction permit for the three cheetahs, one female and two males, had been received, and the import permit had arrived for the female. We had decided to call her Savannah, as cheetahs love the open veld, and we hoped she would love the beautiful plains of grasslands at Thula.

We were preparing to welcome the first cheetah seen in the area since 1941. Ordinarily, I would be jumping for joy, but even this wonderful occasion was dampened by the sadness of Frankie's passing. I struggled to reconcile the conflicting emotions, and thoughts of Frankie haunted me on what should have been the happiest of days. I wished we could have welcomed her in better circumstances, but I also knew that Savannah's arrival would bring much-needed joy.

It was all systems go in the Paarl, in Western Cape, where Chantal and her team were preparing Savannah for her journey. She was darted, and then a DNA sample taken. Vaccines were administered, she was fitted with a tracking collar, and she was ready to be loaded into a crate and then into a small airplane to make the trip to Richards Bay. The excitement from the Thula Thula team was intense as we awaited the arrival of our special new inhabitant. With the experience of Mona and Lisa's relocation still fresh in my mind, I sent two bakkies to fetch Savannah from Richards Bay airport. I was taking no chances with this precious cargo! Vusi was in one bakkie, Christiaan in the other, with Clément and I following in my car. Lynda and Kim completed the convoy. Kim could hardly sit still, she was so excited to be recording this memorable day.

The plane arrived. We got permission to go onto the tarmac once the engines had stopped. I came forward to greet Chantal, who was dressed in rather elegant khaki. Vincent van der Merwe, the cheetah expert, had a big smile on his face as always. He's a man who seems to love his job. And this was a big day for all of us. The crate was unloaded and I peered in to see Savannah. Even in a crate and still groggy from the tranquillizers, she was a beautiful sight. Up close, her spots were pitch black against her tawny coat, and the underneath of her throat a pale cream. I couldn't wait to see this magnificent, agile cat running free, in all her glory.

We lost no time in getting her onto the bakkie and on the

road. A capture and journey is always stressful for wildlife and, in addition, it was a hot day. We followed in convoy. When we arrived at Thula Thula, the bakkie drove immediately to the holding *boma* and reversed in. The men lifted the crate off the back and onto the grass. Chantal and I stood on either side of the crate door.

'One, two, three . . .' she said, and on the final count we raised the sliding door.

'Welcome home,' I whispered.

Savannah came out at a run, her long, thick, white-tipped tail streaking behind her. There was a hushed gasp from the onlookers. Cheetahs are so graceful and so fast, that even a short trot to the other side of the *boma* was magical to behold. Savannah stopped and turned to face us, her golden-brown eyes ringed with the black eye-liner and the characteristic black 'tear marks' running from the inside of the eye down to the outside of her mouth. She looked like a supermodel on a catwalk, turning to show us her best side, giving us a moment to admire her in all her beauty. Kim, Clément and all the rangers were taking pictures and videos. I had invited other photographers too, to record this momentous release, so there was plenty of competition around as to who would capture the perfect shot, the best angle. She made a sound somewhere between a hiss and a soft growl. Cheetahs don't roar like other big cats, but they make all sorts of different noises – they yowl, yip and even purr like a happy house cat.

'This is a dream come true,' I heard Christiaan say quietly behind me. And it was indeed. What a magnificent sight it was – the lithe cheetah in the long grass which swayed gently and glowed gold in the warm afternoon sun.

The cameras whirred and clicked, whirred and clicked.

Savannah stayed in the *boma* for two months, getting to know the smells and sounds of her new home, and awaiting the arrival of the two males who would be her companions, whose import permits we expected very soon. We visited her often, Kim especially. She became completely enraptured by this true beauty of nature and took extraordinary videos and photographs over those two months.

Chantal had told us to work on making a sound that Savannah could recognize and feel safe with, so that when we looked for her in the wild, she would know it was us, and show herself. Christiaan and I worked on a specific whistle. The rangers laughed to see us purse our lips and practise our little tune, which came to mean 'Come out now Savannah, dinner time'. Christiaan fed her every day, whistling the special tune to her as he tossed a piece of meat over the fence into the enclosure. Cheetahs are fussy eaters, and only the most choice and juicy parts of impala were offered to Savannah. She soon came to recognize and respond to his whistle.

It was funny to see her watching the herds of young impala that passed by just metres from her enclosure. She was beautifully camouflaged and hidden in the long grass, and they were totally oblivious to the existence of the predator. I imagined her salivating at the sight of this lunch buffet passing by.

'Don't worry,' I said to her, under my breath. 'You will soon have the opportunity to catch your own lunch!'

In March, we opened the *boma* to set her free. Christiaan whistled his special tune, to encourage her to show herself. Ears pricked to the familiar sound, she came cautiously towards the open gate, looking intently into the bush. 'Come on Savannah,' Kim said quietly. 'Out you come.' He whistled again. The beautiful beast left the *boma* and went out into the world. What a magnificent moment it was for all of us and for conservation. She was a wild and free cheetah now, and soon she would have a mate – in fact two, cheetahs are rather promiscuous creatures – and soon, I hoped, there would be cubs.

That first day, we followed her in a game drive vehicle, at a distance, so that we could record the moment in photos and videos. Instead of going straight to the savannah land in front of the *boma*, she went in the direction of the Safari Lodge, which was close by.

'*Oh non!*' I hissed, smacking the palm of my hand to my forehead.

'What is it?' asked Kim. 'What's wrong?'

'I didn't think to tell the staff at the Lodge that we were releasing Savannah. I didn't think she would come straight here.'

By now, we were pulling into the Lodge. Something moved behind a big Marula tree that shaded the wooden deck. A face appeared, eyes wide and terrified. It was Biyela, the gardener, watching the big spotted cat which had taken a stroll to the suite royale, the honeymoon suite, and flopped down in front of it, her tail twitching.

Biyela needn't have worried, Savannah had no interest in him. She only had eyes for the impala on the other side of the Lodge. A small herd of impala and nyala hang out in the garden where they know they are safe from harm, and where under the admiring gaze of the guests, they can browse in peace, undisturbed.

Until now.

Savannah started walking, not a casual lazy lope, but an intent stalking, low to the ground, her eyes fixed on the prey. She put on a burst of speed, streaking towards the unsuspecting impala. We held our breath in awe, witnessing the fastest animal on earth, in all her glory.

The impala scattered in all directions, making the loud coughing sound that is their alarm call. The call alerts the herd of danger, and lets the predator know that it has been spotted, and that the element of surprise is lost. These impala had never seen a cheetah but they knew enough to recognize that this strange new creature spelled danger to them. The baboons joined in with their own loud barks of alarm. The noise was deafening. As Savannah disappeared into the bush I (rather belatedly) went to inform the Lodge staff of the new arrival. The place was curiously empty.

'Hello? Where's everyone?' I called.

Mabona's head popped up from behind the bar counter, followed by Cindy.

'I came to tell you about the cheetah, but I think you might have seen her . . .'

'Yoh!' said Cindy. 'She was right there, on the lawn. This big spotted cat. We got such a fright!'

'Well, if she had come in, hiding behind the bar wouldn't have helped you,' I said. 'Cheetahs can run and jump.'

There was giggling and nervous shrieking from behind the bar.

'Cheetahs don't attack humans. You don't need to worry,' said Christiaan with a dismissive wave of his hand. 'They don't look for trouble, only food.'

The ladies reluctantly left the safety of the bar.

Biyela wasn't taking any chances. He slowly swept the floor of the Lodge, looking over his shoulder every ten seconds, lest Savannah reappear. He stayed inside the Lodge the rest of the day, sweeping non-stop. Even when every speck of dust was done, he started again, rather than go back to the garden where this new beast might be lurking.

One thing you can rely on in the bush is the efficiency of the 'bush telegraph' or, as we call it in Zululand, 'the Zulu drum'. News of Savannah's release travelled throughout the reserve with lightning speed. When Christiaan returned home to Tented Camp on the other side of the reserve, he found the staff there waiting eagerly – and nervously – for an update on this terrifying new creature who had come to disturb our land of peace and tranquillity.

The truth is that cheetahs are relatively docile. As Christiaan had told the Lodge staff, they don't seek out conflict, especially not with humans. They have never been known to attack a human in the wild, unless they are cornered or injured. It was the antelopes and small game whose peaceful existence would be disturbed by Savannah's arrival.

Life in the wild is a constant cycle of life and death, bringing with it a rollercoaster of emotions. There is so much that we cannot anticipate or control. Within the space of a week, we had lost a matriarch and welcomed a new endangered species. Savannah brought happiness after Frankie's death, just as after Lisa's death, the birth of baby Sissi was a blessing. We would mourn Frankie for a long time to come, but Savannah was a ray of sunlight amongst the dark clouds of loss that enveloped us.

25

Life after Frankie

The Thula Thula herd was thrown into disarray by Frankie's death. Instead of staying together for company and protection, they dispersed, wandering around the reserve. They seemed lost, directionless. There was no sign of the unity that we had come to expect from the elephant family. It was as if, without her leadership, they weren't able to keep the group together.

Elephants are extremely social creatures with deep bonds. Females and younger animals usually stay together in a breeding herd, while older bulls leave and live solitary lives, or in small, loose groups with other males. The Thula Thula elephant dynamic is unusual, in that the males tend to stay more closely attached to the herd. In the weeks when Frankie weakened and died, the group became more separated, less cohesive. The bulls in particular started to wander more, spreading out all over the game reserve.

It was not only the elephants who grieved. The whole of Thula Thula, humans and animals alike, mourned the loss of a great lady. All the staff and even our friends and guests were devastated by Frankie's death. When she heard the news, Kerry Miller, an artist who had stayed with us, made a beautiful oil painting of Frankie, and another of Frankie with Gobisa. I was very touched to receive the magnificent pictures all the way from England.

We weren't ready to lose her. We named the secret little dam that Frankie had found and chosen as her forever home Frankie's Dam, in her memory. It's well known that elephants mourn their family members as we humans do, but they didn't go immediately to her resting place. I wondered, when the elephants dispersed in that strange way, whether they were looking for the dam. We had so many questions about their strange behaviour, including the most important question of all – who would replace Frankie as their leader?

The matriarch is usually one of the older females. She has a long history with the group, and is the repository of deep knowledge about the family, their personalities and relationships, natural resources, predator behaviour and much more. She knows where the marula trees are to be found, and when they fruit. Where on the river they might find water, even in the dry season. She gives her time, care and attention unconditionally to the family members, putting the well-being of the herd before herself.

On 21 February 2021, a month after Frankie's death, the whole elephant family visited us for the first time since Christmas Day, when we noticed that Frankie was missing. This time, they seemed to be heading towards the south of the reserve, where Frankie used to take them when the weather was changing for the worse. They stopped, lingering outside main house. I was delighted to see them together, but deeply saddened by their behaviour.

Then, slowly, ponderously, they walked up and down in front of the fence, Marula at the front, Gobisa at the back, dragging his feet. It was a gait I'd never seen. The teenagers, usually full of life and fun and showing off their party tricks to entertain us, moved as if in slow motion. The elephants seemed sombre. I knew in my heart that what I was witnessing was a funeral march for Frankie. They were coming to share their grief with us. I know some people will say this is co-incidence, or some nonsense that I have dreamed up. But it's none of those things. It's inexplicable, but I've seen and experienced their intuition many times.

The Thula Thula herd came to mark the death of their matriarch, Frankie. Just as when Lawrence passed away and they knew – just knew – that something had happened, and came to visit. Just as, a year later, when we brought the bones of the great bull Numzana to rest with Lawrence's ashes at the dam, all the elephants came to visit the remains of the two great friends and leaders.

From the time of Frankie's death, we watched closely to see how the dynamic of the group changed. I asked Siya to ask the rangers to monitor the herd's behaviour and report back. It was fascinating to watch. They started to gather into two herds. Nana's family stayed together. Frankie's family stayed together. ET and her family moved between the two. The groups would sometimes join up for days or weeks, then drift apart for a while. Everything seemed to be in flux. A new matriarch must emerge, but when? And who?

This was the subject of much discussion amongst us, and the rangers all had firm views and preferences. In a regular office, you might gather round the water cooler discussing favourite football teams, or the outcome of the American election. At Thula Thula, we discussed the new matriarch. We knew it would take a bit of time for the outcome to be revealed, but that didn't stop the speculation. The two most likely candidates were Nandi and Marula.

Andrew was head of team Nandi.

'She'll make a good leader,' he said convincingly. 'She's a bit older, and more experienced. She's learnt a lot from her mother Nana, and I think she has her wisdom.'

'You would say that. Nandi's in love with you!' Muzi teased.

'That's not so . . .'

'It is! She tries to climb into your vehicle!' said Khaya. Everyone laughed – it was true, the two of them had a special bond and we'd all seen how she reacted when they met on a game drive, coming right up close and putting her head into the car, sniffing and nudging with her trunk.

'But what about Nana?' said Muzi. 'Ever since her cataract, Nandi has stuck close to her, taking care of her. She already has a role in the herd. I don't think Nandi will be the one.'

'True, and besides, do you think Nandi has got the temperament for leadership?' I said. 'I worry that Nandi is a big softie.'

'Well you can't say that about Marula!' said Siya. 'I saw her kick Boni's butt the other day. The youngster wasn't crossing the road fast enough and her mother gave her a smack.'

OK, so maybe that's not ideal leadership, but it did show

that Marula expected – and got – discipline and respect from the young elephants. Just like humans, elephants need a combination of assertiveness and kindness. The carrot and the stick. Too much of one or the other, and you don't have a happy herd, family or company!

The other rangers were tending towards Marula, Frankie's daughter, as the likely matriarch, and I have to say, I was in agreement. Although she was smaller and younger than Nandi, and not very experienced, I felt she had the right character for leadership. She and her brother Mabula had their mother's temperament – feisty and firm, entertaining and fun. And she was not entirely inexperienced. In recent years, she had been at her mother Frankie's side, observing and learning the ways of the matriarch.

The herd continued to be unusually fluid. It was only after three full months that Marula emerged as the new matriarch. I will never forget the day that she arrived at the main house to announce herself as their new leader. Gobisa followed her closely, as her protector and mentor, and behind him, the entire herd – Frankie's family and ET's family and Nana's family, united again at last. Marula came to a halt by the fence, right outside the house with her family grazing peacefully around her, just as her mother used to do when she graced us with a visit. It was their first visit since the heartbreaking funeral procession, and the sight of them together, calm and serene, warmed my heart. Although she was young and small, Marula reminded me so much of Frankie that day. She was showing the same strength and leadership.

I hoped that she would remember to stop by for a visit

now and then, as her mother used to quite regularly. Elephants are creatures of habit, and we knew that if the weather started to turn, there was a good chance that Frankie would lead the family to the more sheltered south of the reserve, paying us a courtesy call on her way. Her visits always filled us with joy, and we hoped that Marula would continue the tradition.

The herd has been subtly different under Marula's guidance, they do different things, go to different parts of the reserve. It is early days still, but she seems to be adjusting well to her new role. And so begins a new era in the life of the herd.

26

Problems with Permits

The rangers tracked Savannah daily. I was like an anxious mother. I worried about her out in the bush on her own. When I heard the hyenas calling at night, I said a little prayer for her safety. Cheetahs are small and light compared to other big cats. They are not fighters – they would rather run away than stay and battle. They'll even give up their kills to other animals.

'Has anyone seen Savannah?' I asked Siya.

'No, but we are tracking her by the GPS on her collar. She's been on the move.'

'And we've seen evidence of a kill,' said Khaya. 'So we know that she's catching prey and feeding herself.'

'A lone cheetah is almost impossible to spot, even with the GPS,' said Siya. 'When the males arrive, they'll find each other, I'm sure.'

But when would that be? I asked myself in utter frustration.

We had received the necessary import permit for Savannah, giving permission to bring her from another province, but we were still waiting for the permit to bring the two males from the Free State into KwaZulu-Natal, to Thula Thula. We were stuck in an excruciating administrative bind. We had a beautiful female cheetah in her reproductive prime, of an excellent bloodline, waiting for a mate. We had two perfect candidates in the Free State province, just a few hours' drive away. We had raised the money to transport them to Thula Thula. But the wildlife authorities had not yet granted us the go-ahead.

Three months went by, we waited, we emailed, we phoned, we waited some more. Nothing. The tracking collar that Savannah wore was not working properly, and the fact that we didn't know where she was made us all very anxious. After three months of silence, I received a letter from the wildlife authorities saying that they had done a GPS survey to check the exact size of our fenced boundaries, and their calculations showed we were not big enough for cheetah – we were still not the 5,000 hectares required. In fact, we were only 3,200 hectares.

I was completely astonished. How could this be true? Lawrence had told me that the reserve was 4,000 hectares, and that if you took into account the hilly topography, it was 4,500. This is what I'd always believed and quoted in all our marketing material and interviews. I never questioned it. We were busy with the fencing of the Lavoni expansion, which we thought took us to 5,000 or at least very close.

I consulted Christiaan and Kirsten Youens, the lawyer who helps us with the permits. It turned out that calculating the size of a reserve is not a simple matter. There are different sources – title deeds, aerial surveys, and so on – which do give slightly different results. The way that the area is calculated doesn't take the topography into account, and the calculation was complicated by the fact that our land was made up of a number of different parcels of land, with different owners and title deeds. It was astounding and, frankly, devastating to discover that the land was much smaller than I had thought.

The fact remained that Savannah had plenty of space to roam, and, even more importantly, plenty of food available. When Chantal and Vincent had come to deliver Savannah, they went on a game drive not to see our famous elephants but to check on the food for the cheetah! They came back totally satisfied with what they saw. We had more than enough impalas, nyalas and other small antelopes, to keep Savannah happy for a very long time. So at least I didn't have to worry about that. There was plenty of space and prey for the two male cheetahs, too. But the import permit was not forth-coming.

'What good is one female cheetah, in terms of conserva-tion?' I asked in desperation. I remembered Chantal saying that she considers a release a success if the female gives birth and that offspring is reproducing in reserves.

'Savannah should be breeding now,' said Siya, in agreement. 'But what also worries me is that she is on her own. The hyenas and leopards . . .'

Meanwhile, as we dealt with our land issues and our permit problems, the two male cheetahs were growing into their breeding prime – in quite a troublesome way. The owners sent us videos of these males trying to breed with their mother and sisters. This was shocking. It was clear that we needed to arrange their transfer as a matter of urgency. I sent the video to the people in charge of issuing permits at the wild-life authorities, with a plea to fast-track our application, and let these animals come to Thula Thula where they would be safe, and able to breed productively and healthily.

No answer.

I felt desperate and helpless. We had also applied for a transport permit for Rambo, a big brother for our Thabo, and a mate for our females, seeing as Thabo was clearly not performing his duties in that regard. A week later I received an email from the authorities saying we could not get the permit because Thabo was a troublesome rhino who might fight with the new male, who might then escape and pose a danger to the local communities. I was puzzled by these assumptions.

The real shock came when they told us that if we wanted another rhino, we should get rid of Thabo and relocate him to another reserve. We were appalled at the idea of removing Thabo from Thula Thula, the only home he had ever known. He had arrived as a traumatized orphan at just one day old, having witnessed the brutal death of his mother at the hands of poachers. Was he now going to lose the adoptive family who had loved and cared for him for twelve years? As well as Ntombi, his love and companion?

Leaving his territory would be completely traumatic for him. Let's face it, Thabo owned Thula. Thabo was a legend! He was known all over the world for his history, his love for humans, his bad-boy antics. He was a unique rhino, with an incredible life story, and he belonged here, in his home turf. Every family has a problem child or a troubled teen at some time. What, must they send them away? Besides, who would want a so-called 'problem' rhino? We knew and loved Thabo, and we knew how to handle him. We also knew that there was indeed a possibility that there would be conflict between Thabo and his big brother. Wild animals do fight; it's part of life in the wild. But Thabo had never been aggressive with other animals. He bowed to the superior might of the elephants.

Rhinos are facing extinction. In our efforts to understand and help Thabo, we looked into how orphan rhinos fare in adulthood. I was surprised to discover that orphaned male rhinos often fail to reproduce successfully. No one is quite sure why, but it seems that there is something about the early trauma or lack of role models that seem to render them unwilling or incapable of reproducing. No matter how well we look after our little rhinos, there are things that humans can't teach them, and when they go out into the big wide world, they don't do well in this regard. The long-term effects of poaching are truly tragic. Thabo had the additional challenge of being the only male on the reserve. We hoped that having another male around competing for females might spur him into action! We have two breeding-age females without a male who could impregnate them. It seemed like common sense to get a male rhino as soon as possible and start making babies. There was nothing we could do, other than reapply. Which we did.

My problems were about to get bigger.

In June 2021, a threatening letter was delivered to my door, accusing me of non-compliance, of lying about the size of the game reserve (an infringement that could result in a fine or even a prison sentence!) and of breaking the law. The wildlife authorities threatened to remove Savannah. This would mean darting her again, administering an anaesthetic which always carries a risk to the animal, and is particularly risky for cheetahs. Not to mention the unnecessary stress on Savannah.

How is that in the animal's best interests?

I was beyond devastated by this latest turn of events. In a hugely difficult year, in which we had fought to survive for sixteen months of Covid restrictions, and we had lost our beloved Frankie, this was a low point. The problems were just piling up. I hardly slept for days, worrying about Savannah's safety, about getting the male cheetahs, about the land, about my legal and permit problems.

That evening, I phoned Clément in Durban.

'I don't know how much longer I can fight like this,' I told him, biting back tears. 'I only want to do what's best for the animals, for conservation. That's why I do what I do, every hour of every day.'

'I know, *chérie*,' he said. 'This letter, it's horrible . . .'

'I understand the need for rules and regulations. I've made my mistakes. Maybe I'm not the very best with admin, but surely someone could have phoned me and talked to me. Not send me this.'

I started weeping hysterically. I honestly felt that I was cracking up from the stress. My dogs, who were lying comfortably on the couch next to me and spread at my feet, inched closer. They always knew when I needed comfort.

My darling little Gypsy crept onto my lap and licked my tears.

'Try and get some sleep,' Clément said, doing his best to soothe me from miles away. 'Tomorrow is another day.'

Perhaps, but it seemed that each day was just as difficult. Clément always has the ability to find the right words to make me feel better, but this time I was too devastated and overwhelmed to take comfort. I was worn down by the succession of defeats and sorrows. I just felt like running away from the insanity of it all, away from the letters and lawyers and disappointments, to just be alone, to gather my strength and my thoughts. I was at rock bottom. I don't think I'd been this downhearted since I lost Lawrence.

And then a miracle happened.

27

Bigger, Better and Brighter

The day after I fell into such deep despair, I awoke feeling somewhat better. There will be no more dwelling in self pity, I said sternly to my reflection in the bathroom mirror. I am putting that behind me and moving on.

I picked up the phone. I messaged Christiaan and Lynda and asked for a meeting. Now. This morning. They arrived, not knowing what to expect.

'Right!' I said, forcefully, hands on hips. 'We are looking to the future. Right here, right now, we are going to find solutions to our problems.' They looked pleased to see me back in positive, problem-solving form.

'Starting with the land issue,' I continued, tapping the map of Thula Thula that hangs on my wall, showing the roads, fences, rivers and boundaries, and the newly fenced Lavoni section outlined in green felt tip.

I looked at the map I'd seen a thousand times before, but I saw something new – a game reserve across the road from Lavoni, well established, properly fenced. I knew it had recently been sold. 'Here's an idea,' I said.

There and then I called Michael Crichton, the new owner

Jo Malone visits Thula Thula.

Christmas with lots of gifts from our generous friend Jo Malone.

Lucy as a baby. We found her at the entrance of the game reserve.

Christiaan and his beloved dog Bruce.

Surrounded by my lovely dogs.

Daisy the meerkat.

Mona and Lisa getting dehorned.

Feeding the wildlife community during Covid. It was a scary time for everyone but a real privilege to be able to help.

Thabo attacking the trucks during road construction.

Our vet Trever Viljoen doing Frankie's check-up – he's taking blood samples and injecting vitamins and antibiotics boosters.

Feeding Frankie with Andrew.

The Dube Ridge expansion signing day. A really special moment for the community and wildlife conservation.

The fencing of Lavoni with the volunteers during lockdown. It is Lavoni land in the background.

Elephants exploring the new land at Lavoni.

The staff of the Tented Camp and the Safari Lodge
all dressed up for a special occasion.

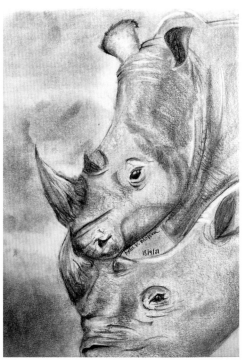

(*top left*) A fertility operation for Ntombi and Thabo.
(*top right*) A drawing of Thabo and Ntombi from a
young conservationist named Ayan.

Rambo and Thabo's first encounter.

of Zulweni, the game reserve over the road. He was friendly on the phone, pleased to hear from me and very happy to meet.

'I'm in Durban for the week,' he said, 'but I'll be coming back to Zululand at the weekend. Let's make a plan to meet up next week sometime.'

He did not know how impatient I can be when in the grip of a good idea! I was all fired up by then, and as far as I was concerned, there was no time like the present.

'I'll come down to Durban tomorrow,' I said, rather to his surprise. 'What time suits you?'

The very next day, I was sitting in a coffee shop with Michael. He was a successful businessman who had decided to buy a game reserve and dedicate his life to conservation. Rather like Lawrence and me, who had left the big city life to come to the Zululand bush. As we talked, it was clear that we shared a passion for wildlife conservation and a bold vision for the future of wildlife and tourism in Zululand. I asked about his plans for his reserve, and he spoke with quiet passion about his 1,200 hectares of beautiful Zululand bushveld and the residential development he wanted to build on it. I liked him immediately. He was soft spoken and thoughtful, with a calm, gentlemanly manner.

I quickly came to the point. 'Michael, what would you say about joining our two properties?' I asked. 'Thula Thula is well-known and well-regarded, and we have rhinos and elephants that you don't have. It would be beneficial to you and everyone who owns a home at Zulweni to be joined with us and have these animals on your land. We need access to more land for the benefit of our elephants and other animals, and to meet the space requirement to solve our permit problems. What do you say?'

'I like it,' he said. 'I think it would work well for both of us, and for the wildlife.'

We made a deal, there and then. It was simple and perfect, a win-win. With our 3,200 hectares and Michael's 1,200, we had 4,400 hectares. We were within sniffing distance of the elusive 5,000 hectares.

'It's been a pleasure doing business with you,' he said, smiling and shaking my hand.

'Likewise,' I said. 'I'm delighted!'

'When I'm back in Zululand, we'll finalize the details on the project,' he said.

I walked back to the car with a giant grin on my face and a sense of excitement and possibility. Twenty-four hours ago I'd been in the depths of despair, and now I felt more hopeful than I had in weeks, maybe months.

He was as good as his word. Within three weeks the deal was done.

There were still a few details to work out – like how the animals would move between the two properties, given that there was a road between us. It is important that the animals roam, mix and interbreed, to diversify the gene pool and lead to healthier, stronger wildlife. On a small, fenced reserve, the breeding populations are small and stable, and the animals can become inbred. This can lead to problems, particularly for the larger antelopes, buffalo, and wildebeest. Fortunately, there was a three-metre underpass under the recently built main road. Michael, Vusi and Christiaan went down to look at the tunnel.

'It's big enough for the elephants to get through,' Vusi reported back. 'But not the giraffes.'

This was good news. The animals could pop through the tunnel, under the road, and find new mates. Except for the giraffe, who were too tall to fit through the tunnel. No romantic visits to the neighbours for them, sadly! I was still glowing from this stroke of incredible good fortune when another one appeared.

Just three days after I met with Michael, Larry called, saying, 'Françoise, I want to bring the leaders of the Ubizo community to meet you. They are the new owners of the Dube Ridge land on the southern border of Thula Thula. They have received 1,100 hectares in a land restitution claim and they would like to join Thula Thula.'

Two days later, I met Thomas Cebekhulu, the chairperson, and other representatives of the Ubizo Communal Property Association. They were the most charming people. Just as with Michael, we understood each other, and saw how this could be a mutually beneficial relationship. They were eager to create a traditional Zulu village on Dube Ridge, for tourists to visit. They can learn about Zulu cultures and traditions, and hear locals tell stories of the fascinating history of the Zulu nation and the great King Shaka. Being connected to Thula Thula would benefit them, and bring guests to their tourist village, and create employment opportunities for their people. We, of course, were keen to have more land. Within half an hour we'd reached an agreement.

We had done it! With our two new partners, we had our 3,200 hectares plus Zulweni's 1,200 hectares plus Dube Ridge at 1,100 hectares, which equalled 5,500 hectares.

We had the required 5,000 hectares and more. This was a truly historic moment. With the joining of these new lands, the Greater Zululand Wildlife Conservancy was born. I knew that this was only the beginning of great things for all of us.

The only concern was that the new land at Dube Ridge still had to be fenced, and that would cost money. But that was a problem for another day. For now, I was on top of the world!

After the discouragement, the agonizing wait for permits, the shocking news about the amount of land, there was a little light at the end of the tunnel. I felt sure that it would only get bigger and brighter. I have always believed in my good friend karma, and she came through for me once again. In my darkest hour, when I was ready to give up, two un-believable life-changing, win-win opportunities had fallen from the sky.

We had risen from the ashes and achieved the impossible. With our wonderful new partners, we were bigger, better and stronger than ever. And this was only the beginning. There was still the other 2,000 hectares of community land that had been approved in principle but had not been finalized because of the pandemic. When that went ahead, it would take us to 7,500 hectares – almost twice the acreage it had been before we had joined forces with these excellent partners. The Greater Zululand Wildlife Conservancy will be the largest game reserve in the area, and the closest to Durban.

I felt Lawrence smiling over me that day. This was his dream coming true.

And then came another stroke of immense good fortune.

A few weeks after all the excitement of the expansion plans, a Cape Town couple called James and Laurel came to stay at the Tented Camp. Laurel had read *An Elephant in My Kitchen*, and Lawrence's *The Elephant Whisperer*, and had come to Thula Thula to discover our little piece of Zululand paradise for herself and to meet our famous elephant herd. Well, they fell in love with our elephants, and talked with great animation and emotion about the amazing experiences they had at Thula Thula, thanks to our wonderful game rangers. We chatted one evening, and I discovered that as well as being animal-lovers, they were investors in the movie being made of Lawrence's book. They showed so much interest in everything to do with the reserve that Christiaan told them in some detail about the exciting expansion project, and how our next step was the fencing of the new land that belonged to the local community. James started reading *An Elephant in My Kitchen* when he was here.

The following week I received an email from James asking more questions about the expansion and the community land. I explained about the importance of increasing the habitat for our elephant herd, and I mentioned the fencing project that was crucial to this new step. I told him about the community's plans for their participation in the tourist economy and the jobs they hoped would come from that. I soon received another email.

Dear Françoise, I have just finished your book. Having

read about and seen the great work you are all doing in conservation, we would like to support your conservation fund. Please send us the account details.

I thanked him warmly and did as requested. Any donation is received with gratitude and put to good use – the need is endless, even more so during the pandemic. When that donation popped into the account, well, let's just say that it was very generous. The funds gave me breathing space without the constant financial worry. It meant that as soon as all the paperwork was in place, we could start expansion immediately. It was a dream come true.

After months and months of stress and total discouragement, I had been at such a low point in my life that I wanted to give up and run away – and now look! In the space of two months we had secured an extra 2,300 hectares, taking us to 5,500 hectares. We had close to doubled in size. What's more, we had the money to fence it. Best of all, we were 'legit' at last! We were big enough to keep elephants and get our cheetahs. All my problems had been solved, and we were growing big and beautiful with our partners and new friends.

Isn't that one of the wonders of life? That at the most unexpected moments, in your darkest hours when you are facing the greatest adversity, something wonderful can happen. In fact, I often think that when your back is against the wall, when you are out of options, that's when you are pushed to be at your strongest and most creative.

The pall of disappointment that had settled over us all dissolved like the early morning mist over Mukulu Dam. We snapped into action and started working on the new management plan for the next phase of expansion of Thula Thula. My troubles were over. Or so I thought.

In fact, worse was soon to come.

28

Notes from a Burning Country

On Thursday 8 July 2021, we heard news of civil unrest in our province of KwaZulu-Natal. The protests were sparked by the imprisonment of South Africa's ex-president, Jacob Zuma, who had been sentenced to fifteen months in jail for contempt of court. Zuma's home is in the heart of Zululand, not very far from Thula Thula. As a Zulu elder, he remains popular amongst certain sectors in KwaZulu-Natal, despite his tarnished reputation and the corruption charges against him (he denies any wrongdoing). His supporters set up roadblocks on the main highway between Durban and the industrial hubs of Johannesburg and Cape Town.

Unrest is concerning, of course, but it tends to blow over quite quickly. Besides, we live in a quiet backwater here in the bush, miles from the city. To be honest, I didn't take it too seriously at first. It soon became clear that this wasn't some flash of anger that would be over in a day or two. The next day brought the news that the violence and looting had spread within KwaZulu-Natal, and to the province of Gauteng, to Johannesburg, the country's largest city and economic hub.

Work came to a halt in the office as we watched television footage of the unfolding chaos in horror. We could hardly believe what we were seeing. Huge mobs ransacked shopping malls and centres, smashing windows and looting food, electronics, clothes and liquor. They pushed, shoved, shouted and emerged triumphant with whatever they could lay their hands on, including fridges! Others attacked and burned warehouses and factories in industrial areas in furious acts of destruction. The scale of the devastation within our beautiful province was shocking. We watched in horror – the flames, the anger, the violence. The carnage was like something from a movie – chaotic, apocalyptic, and terrifying.

And then Aphi pointed at the screen and shouted, 'That's Empangeni! That petrol station, look!'

'She's right. It's the one on the main road,' said Portia.

True enough, it was. We were watching footage of our local town burning. It felt utterly unreal. The staff were distraught. This is the town where we shop, where their friends and relatives live and work, or their children go to school. And now the shops were being invaded, petrol stations set on fire. Cars stoned and burned. I felt helpless, watching this anarchy and destruction, the looters having a field day, with no police or army in sight.

My immediate concern was a party of guests – children amongst them – who were due to leave on Monday and drive the two hours to Durban Airport to fly home. Larry, our head of security, called early on Monday morning to say that the toll booth near the airport had been burned and crowds had barricaded the entrance to the airport. Our guests would not be flying today. We changed their flight to the next day, hoping that the situation would have calmed down and they could leave safely.

We sent the guests off very early Tuesday morning in a convoy. We were in constant contact with Larry, who was circling the area by helicopter, informing us of the no go areas. It was extremely stressful, but thankfully, the guests made it out safely.

I made myself a cup of tea to calm my nerves and enjoy a moment's relief at their safe passage, but before I had even finished the tea, I got a call from Larry.

'Françoise, we've heard from informers that Thula Thula is being targeted by the rioters and looters.'

'What do you mean? Who would riot and loot at a game farm? We have nothing they want.'

'I can't answer that, but this is a serious threat, Françoise. These are reliable informers, and they say that the plan is to come to the village . . .'

'Our village? The rioters are coming to Buchanana?'

'Yes. My informants say they plan to loot and burn China Mall.'

Our tiny local shopping centre, supplying the whole village with food and other necessities, is run by Chinese people. The only other shop in the village is a fruit and vegetable stand which belongs to a local villager. The Chinese shopkeepers had started with one little shop and now had a small empire of three little shops, which we call China Mall. I always thought how strange it was to see Chinese men and women speaking fluent Zulu in a tiny village in Zululand.

'Oh no!' I said, thinking of the poor owners who had built up this small business so far from home.

'And my sources say . . .' Larry hesitated a moment, and then continued. 'They say that then they'll come to Thula Thula and burn the Lodge and the main house.'

'When?'

'This afternoon.'

'This afternoon?' I repeated, in disbelief. My heart was racing. How could this be happening? 'What about the police? And the army?'

'They've got their hands full. They're not going to help us. But don't panic. I'm organizing men and vehicles. We're going to come and defend Thula Thula.'

Defend Thula Thula? What movie had I found myself in? Would Bruce Willis be arriving next?

I was in a total panic when I got off the phone. If the police couldn't help us, we would have to rely on ourselves and our community. Christiaan and Vusi went to a meeting with another game reserve down the road to discuss how we and our neighbours could work together to protect ourselves.

You've got to love the can-do spirit of local South African farmers. The Farmers' Association snapped into action and immediately started creating a new back road with a bulldozer, so that we could reach the main road, if our entrances were blocked and we had to evacuate.

'We also have two helicopters, in case the roads are impass-able,' said one of the farmers. 'We can get you out by air if necessary.'

I was grateful that they were so organized, but in my head, I was saying, *Evacuate by helicopter? Where will I put my twenty-eight elephants? My nine dogs? My four rhinos? Let alone my fifty staff*? I knew that evacuation was not an option for me. I could never go and leave them behind. I would have to stay here, whatever happened.

Larry arranged for heavy firepower and trained men, but this sounded like a situation where we needed all the help we could get. I asked who at Thula Thula knew how to shoot. Christiaan and most of the local farmers were part of the generation who had done compulsory military service in South Africa and were comfortable with firearms. We canvassed our neighbours to see who might be able to help us with additional weapons. Once again, I had this sense of unreality – what, now I was mustering men to defend our little piece of paradise in the bush? With guns?

It would be a stretch to say that I could shoot, but I did take shooting lessons in the 1990s. There was a great deal of fear and unrest in those years before the democratic election in 1994, and Zululand was particularly affected by conflict. Lawrence bought me a little gun so I could protect myself if necessary, but I needed to get a firearm licence, which meant shooting lessons. Those poor instructors, I hope they've recovered from the experience of teaching that mad French woman to shoot! Like any southern French person, I talk with my hands, gesticulating wildly. That's not such a good idea when you are holding a gun. Everyone cowered and ducked for cover whenever I started speaking, knowing about my tendency to wave it about to illustrate some important point in my story. Another time they put a bazooka on my shoulder and I turned around to ask the instructor where to press to shoot. Oblivious to the fact that I was aiming the massive firearm straight at them, I wondered, why are these guys diving to the ground?

I managed to get my gun licence without mishap. I'm sure there was a sigh of relief from the instructors! Fortunately, I never had to use the gun. If the situation had ever presented itself, I would have had to rummage around at the bottom of my bag to find it, and then fumble about trying to remember how to unlock the safety, and by that time, well, it would probably have been too late to do any good. For a few years I was armed and dangerous, and then I gave the gun away when the government instructed all SA citizens to disarm ourselves to reduce gun violence.

I rather wished I had that little gun now.

I called a meeting of our staff and told them what we'd heard about the planned attack that afternoon. Of course they were horrified at the thought that someone would threaten us, and threaten their jobs and livelihoods.

'Please, call your friends and family in the village,' I asked them. 'Tell them about these threats to Thula Thula. Get their support.'

They all got on their phones, telling everyone they knew not to support any violence or destruction, and to tell any mischief-makers to leave us be.

Nothing happened that afternoon. We will never know whether our friends in the village convinced them, or whether word went around that Larry's highly trained security personnel were protecting us and the would-be assailants got cold feet. That night, the panic and stress kept me up. I could hear the drums in the village, but had no idea what they meant or what was going on. I didn't sleep a wink.

Fortunately, we were untouched at Thula Thula, but the devastation elsewhere was shocking to behold. Malls, shops, factories, schools and warehouses were attacked. Parts of Durban looked like a warzone, with burnt-out buildings and broken glass. The looting was massive, both by poor people grabbing food, and by shameless opportunists taking whatever they could and leaving a trail of destruction. There was a video doing the rounds of a guy wrestling a massive flat screen TV into the back of his car (unsuccessfully, it turned out).

'That's what I would look like getting my Mabula into the evacuation helicopter,' I said to Lynda, showing her the video on my computer. Even in the worst time, you have to have a laugh.

It later emerged that what had happened was in fact an insurrection. There were agitators leading and encouraging those crowds, intent on bringing our province and our country to its knees.

Well, they did not succeed.

29

The Great Spirit of Ubuntu

When the looting and violence had started, the response from ordinary South Africans was extraordinary. Communities came together to defend their own neighbourhoods. Clément was in our apartment in Umdloti, north of Durban, where the whole community rallied to create barricades to block access to the village from the main road to keep intruders out. Clément did two hours of community protection every night in the winter cold. He had been in the army by the Angolan border, and said that the patrolling reminded him of his youth. Clément had his gun, but others were armed with whatever they could find, from shotguns, to pangas, to cricket bats.

With roads blocked and businesses closed, food and petrol were soon in short supply. Clément baked bread to bring to his new vigilante friends, along with hot coffee with a tot of Amarula – a deliciously sweet and creamy local liqueur made from the fruit of the African marula tree, much adored by elephants.

'For courage, Françoise,' he said. 'Just a nip. We are like the legionnaires who had a tot of cognac before a battle.'

Mabona and I rationed the food we had so that everyone

was taken care of, and our supplies would last. 'You know what? We eat too much anyway,' I said. We all patted our tummies rather guiltily at that. There was enough to go round, we shared what we had and no one went hungry.

Every community along the coast united in great spirit to defend their shopping centres, homes and properties. In our local town of Empangeni, a human chain encircled Five Ways Mall, the biggest shopping centre in the town, to protect it from looters. If not for this brave community, it would have been lost.

We only realized later just what incredible work ordinary people did at that time. The TV news showed only violence and destruction, spreading even more fear and anguish to the world. I wish the media had shown how these little islands were kept safe against looters. I would have liked everyone to see the unsung heroes, individuals and communities who bravely worked together to save buildings and properties and protect lives.

The manager of a big supermarket in a shopping mall poured cooking oil all over the floor outside the store. The looters found themselves slipping and sliding and falling on the slippery tiled floor. His clever action kept the store from being ransacked, and the video made for an amusing interlude amongst the horror.

The looting and destruction continued for a week. Then the clean-up began. It was a massive job. The streets were littered with broken glass, packaging, dirt and rubble. Pillaged loot had been dropped, and food items had started to rot in the streets. But here again, communities rallied together to clean up their roads and shopping malls. People went out with their brooms and rakes, and when their own streets and suburbs were clean, they went to help in other neighbourhoods, determined to repair and rebuild what had been damaged and destroyed.

Trucks arrived from Cape Town and Johannesburg, filled with donations of food and other essentials for people who had lost everything, or couldn't feed their families. I've never been as proud of South Africans as I was in the days after the riots. Their courage and the spirit of *ubuntu* was in evidence everywhere – people helped each other, showed unity and courage and kept a positive attitude.

Because I live with elephants, I tend to compare them to humans. Elephants are said to be destructive. In fact, they are sometimes called 'demolition artists', because they break things down. A better name for them might be 'landscape artists' or 'environmental engineers'. For millions of years, they have shaped the world around them. Ecologists call elephants a 'keystone species', which means they are critically important for the survival of an ecosystem. Without its keystone species, the entire ecosystems would be dramatically different or even cease to exist.

Elephants' destruction, unlike humans' destruction, has a purpose: it creates life. When a tree is pushed over, a gap is created in the canopy. Smaller, younger trees have the opportunity to grow and there are trees of different heights available for browsers like kudus, nyalas, and bushbuck to eat. When they move about, they trample dense bush or forest, creating pathways and making room for other animals to live. Like giant gardeners, they prune scraggly trees when they munch on them, encouraging them to grow thicker and more sturdy. The way elephants feed also helps ensure that no one plant species dominates the environment.

They use their tusks to dig for water in drought years, incidentally creating watering holes for other animals to access fresh water. Elephants eat a huge amount, and what goes in must come out, right? Elephants have quite inefficient guts, so their dung contains a lot of semi-digested leaves, grass, bark and fruit. Some plant species have even evolved to have seeds that need to pass through an elephant's digestive tract before they can germinate. In this way, elephant dung distributes plant seed over great distances, and adds nutrients to the soil.

Elephants affect the whole ecosystem, the plants, the land and the water, which means that they also impact other animals – from the big predators that prey on the herbivores that graze on the grassland the elephants have cleared, right down to the hard-working dung beetles who live in the elephant droppings.

So, you can imagine why I get cross when I hear people talk about how destructive elephants are. Their big feet and big appetites do nowhere near the damage that we humans do through climate change, mining and the destruction of the wild bush land and the fragile ecosystems that depend on it.

Elephants can damage land or property if they are in a confined space, or when humans and elephants live close to each other, but an elephant is never wantonly destructive. There is no comparison between an elephant trampling the bush, and the worst of destructive human behaviour.

Protestors set fire to a chemical factory in Durban during the riots. It created an ecological catastrophe when water used to put out the fire washed toxic chemicals into water spilling into the nearby river, and then the sea. The beaches north of Durban were littered with dead fish and crayfish. It was horrifying to see. Miles of beaches were closed for weeks. We still don't know what the long-term consequences will be.

But I remain positive. Mother nature teaches us again and again that damage isn't permanent, that very often, after a fire, a hailstorm, or a harsh pruning, regrowth is stronger and more beautiful.

In our worst times, we witnessed the amazing positive attitude and incredible team spirit of our Rainbow Nation, South Africans from all walks of life, standing together to rebuild the country for a better future. The way people rallied and pulled together reminded me of our elephants, who live and work together in unity, giving each other support and comfort. I believe that we can learn from this experience and go forward even stronger.

In the human world, as in the elephant world, unity is power.

Or, as we say in Zulu – *Ubuntu.*

I am, because you are.

30

The Silent Extinction

While the world was focused on the pandemic, which was now in its second year, elephants moved one step closer to extinction. The International Union for Conservation of Nature (IUCN) Red List of Threatened Species listed the African Savannah Elephant as endangered, and the African Forest Elephant as critically endangered. They had previously been assessed together and classified as vulnerable, and this new classification means that they are considered at high risk of being declared extinct in the wild.

Between habitat loss and poaching, there is a real possibility that we – or at least our children – could be living in a world without elephants. This news horrified me, and also galvanized me to work even harder to take care of our elephant herd, and expand the land available for them.

We can't talk about endangered animals without talking about rhinos, which are critically endangered, on the very brink of extinction. The latest State of Rhino report, published by the International Rhino Foundation, says that the estimated rhino population in Africa is about 18,000 – a 12 per cent decline in the past decade. The South Africa white rhino population has plummeted by more than two thirds in eight years. Poaching declined briefly in 2020, when borders were closed as a result of the Covid pandemic, but poaching incidents are again on the rise.

While the plight of elephants and rhinos is well known, other African animals are facing possible extinction and getting little attention. One animal that's less high-profile (if you'll excuse the pun) is our tallest mammal, the giraffe. They are in real trouble – their numbers have fallen by 40 per cent in the last three decades, and there are now only 68,000 of them. Conservationists are calling it 'the silent extinction', because there's so little awareness of their situation.

Giraffes are the most beautiful, graceful animals with their long, elegant necks and sloping backs. Up close, you can admire their intricate patchwork markings, each spotted pattern unique and different, like human fingerprints. (The same goes for zebra, by the way – each stripe pattern is unique. And in case you wondered, zebras are black with white stripes, not white with black. The skin is black beneath the fur). Coming back to the giraffes, I can quite see why the Greeks believed these strange looking fellows to be a cross between a camel and a leopard (which explains their species name, camelopardalis). Aside from their long necks and their great height, giraffes have many extraordinary features. Their tongues are a dark purplish colour. They nap in spurts of a minute or two, while standing up. They seldom lie down, because as prey animals it makes them vulnerable – getting up is not easy when you're built like a set of scaffolding!

These unusual and elegant animals are hunted for their

meat, skins, bones, hair, and tails. I can't imagine why anyone would want to see their beautiful spotted pelts on a wall or a floor for decoration, but they do. Giraffe body parts are also used for *muthi* (traditional medicine), and their bone marrow is believed to cure HIV. Their black-topped fly-whisk tails are coveted as status symbols and the long wiry black hairs used for making bracelets and other items. It was only in 2019 that trading in their body parts was specifically regulated under the Convention on International Trade in Endangered Species of Wild Fauna and Flora (CITES).

We have eighty giraffes at Thula Thula, and they are a great favourite with visitors who always oooh and aaah at these statuesque beauties. In silhouette against the African sunset, a group of giraffes – called a 'tower' or a 'journey', isn't that lovely? – is something magnificent to behold. They are almost as irresistible to the photographers as our elephants or rhinos.

Giraffes don't have quite the same strikingly individual personalities as our elephants, but there are a few who stand out. The rangers named one young giraffe Khabila, after the chap who does maintenance on the Lodge. Only later did they tell me why – the animal has long skinny legs and knock knees, like his namesake! Once the man discovered that his name had been appropriated for the giraffe and why, the poor guy stopped wearing shorts around the place, and now does his work in trousers no matter how sweltering our African summer days.

We had two very elderly giraffes, ancient fellows who would undoubtedly have gone to meet their maker years ago if there were lions at Thula Thula. Shaka is so old and doddery that he can't even walk straight, and his pelt has got darker with each passing year to the point where his patches are the colour of dark chocolate. The other old man of the group, Uneven – so called because of his uneven horns, one longer than the other – was almost equally decrepit, but he was getting paler by the year, his patches the colour of light golden honey.

'Look at them, the two *mdalas* (old men) limping along,' said Muzi one day, some months back, as the herd crossed the savannah land in front of my house, on their way to the little Gwala Gwala Dam for a drink. 'Have you noticed that they are never together? Always on opposite sides of the group. They've never been friends, these two old ones. I think they are competing.'

I hadn't noticed until he pointed it out, but if Muzi was right, and the two old chaps were in some sort of competition for Number One Ancient Giraffe, Shaka finally emerged the winner. I was driving from my house to Tented Camp when I saw something unusual – all of the giraffes were gathered in a tight huddle, staring at the ground. I had never seen anything like it. Although they live in groups, giraffes don't have very strong social ties (other than mothers and babies); they tend to be rather more spread out. What could they be looking at? I couldn't see through the long grass, lush after the summer rains.

I was pondering this strange behaviour when Muzi arrived, bringing guests on a game drive. He stopped his vehicle alongside mine.

'Look at the giraffes,' I said, after I'd greeted the visitors.

'Why are they all huddled together like that looking down? Isn't that strange behaviour?'

'Uneven has died. I saw his body there this morning. They are gathered around his body. Paying their respects to the old *mdala.*'

This was something I'd never seen before or heard of. Elephants are known to mark the death of one of their herd. They visit the remains of dead family members, stroke their bones and even rock back and forth in what seems to be a mourning ritual – we have seen that Frankie's bones have been moved around by the other elephants. A grieving elephant may withdraw from the social life of the group, and their sleeping and eating may even be disturbed, just as it is with us humans when we are very sad. But giraffes? Muzi said that it's a rare sight but that it was not unheard of for giraffes to mourn their dead in this way.

As time went by, and Uneven's corpse decayed, the rangers observed another interesting phenomenon – the giraffes appeared to be eating the bones. Now that was really strange – giraffes are herbivores.

'It's called osteophagia, which means to eat bones,' said Christiaan. 'They don't really eat them, they lick and chew on them. Giraffes' skeletons are so big and their bones so long that they need a lot of calcium and phosphorus.'

Well, I learned something that day. One thing's for sure – there's always something new to discover in the bush. Even after twenty years!

Giraffes only have one calf at a time, and they are pregnant for fifteen months, which means they are not exactly prolific breeders. The little ones are up and on their feet within an hour, but they are very vulnerable to predators in the first few months – in the wild, as many as half of them are lost. But here at Thula Thula they breed well and, because they have no predators and there's no hunting on the reserve, their numbers grow and grow. One of our lockdown gifts, along with Jo the hippo and Sissi the rhino, was June the giraffe, who was born in – you guessed it – June. Such a cute little thing she was, a perfect tiny replica of her mum, wandering beside her on her wobbly legs, batting her long eyelashes.

We had rather too many young male giraffes at Thula Thula. Ideally, you only need one big male to a few females, so to balance the herd and control numbers we wanted to relocate some of them. Fortunately, other reserves were in need of more bulls and eager to introduce new genes to their populations, so we decided to give some of ours away.

Catching and moving giraffes is not a DIY operation, so we employed a professional game capture company to do the job – ably assisted by our own team, of course. Step one in any game capture operation is to find the animals! Vusi and some of the guys from the capture company jumped into two bakkies and went off in search of giraffes. Specifically, they were looking for young males. Smaller animals are easier and cheaper to capture and relocate. They found some giraffes, but the young males were mysteriously absent.

Animals have a very strong and strange intuition and it often happens that if you are determinedly looking for a particular animal, they seem to be just as determinedly avoiding you. When we applied for the permit to get a cheetah, we were asked to send photos of the herds of small buck and antelope, to show that there was plenty of prey available for the big cat.

'That'll be easy,' said Kim, heading out with her camera. 'I'll be back in an hour or so.'

There are plenty of impala and nyala around Thula Thula; you don't need to look very hard to find them. Except for that day. Kim drove around and around for hours, spotting one lone nyala here and another couple of impala there. Certainly nothing resembling a herd.

'I'm telling you, they were hiding from me,' she said grumpily, coming home hours later, hot and tired after her fruitless mission.

The same thing happened with the giraffes on capture day. There were females aplenty, and big old bulls. But not a young male in sight! The rangers finally tracked down a herd with males and females and got to work. The helicopter moved in and the vet darted six giraffes from the air. A light sedative was used so that the giraffes stayed mostly on their feet.

'Because of their height and their long necks, blood pressure is an issue when they are on the ground for any length of time,' one of the capture guys explained. 'They can have a heart attack. They are also difficult to move if they're completely anaesthetized. This way, they can still walk.' Everything is planned and executed to cause the least possible stress to the animal, and the least risk to them and the capture team.

Within minutes, the sedated giraffes were showing signs of slowing down. Once an animal was reasonably docile, the capture team moved in with ropes – the primary tools for giraffe wrangling. I watched a giraffe sink to its knees. Six or seven guys ran to him, one slipping a halter over his head, a couple more putting on a blindfold. As he got unsteadily to his feet, the others slung the ropes around him, attaching them to the halter and around his hindquarters. They were like dancers around a maypole.

Once the animal is blindfolded, and the ropes are tense, it doesn't struggle too wildly and it can be guided to the trailer. It's quite a task, even with a small giraffe, and needs a big team of rangers and helpers to pull the ropes – while keeping clear of the powerful neck and the long legs, which could give you a kick, possibly a fatal one.

While they wrangled the giraffes, the rest of us shouted helpful advice from a distance:

'Mind the feet!'

'Watch out Vusi!'

'Bring him this way! Left, left . . .'

Even in their sedated state, the giraffes were not exactly compliant – one chap splayed and straightened his front legs and braced himself to avoid taking another step! But one by one the six giraffes were lassoed and loaded onto the trailers.

The rest of the herd stood a little way off gazing at this spectacle. Giraffes are known as the hippies of the bush. They are quite chilled, they move slowly, and they are not necessarily the most wide awake or the sharpest animals about. I would have thought they'd make a hasty exit when they saw their fellows being roped up and shipped off, but they stood around watching in what looked like mild bemusement.

'Hey guys, Gerry looks a bit unsteady on his feet, doesn't he?'

'What's that human doing with the rope?'

'Oh, you know humans. Mysterious animals.'

'Hey, Jacob is going in the huge car.'

'Never been in one. You?'

'Nah.'

'Bye then Jacob, see you around Gerry.'

The captured giraffes were taken to the airstrip – or Thula Thula International Airport, as we like to call it – and transferred to big trucks that were waiting there to take them to their new home. They did look odd, their necks and heads sticking out of the top of the tall containers on the back of the trucks as they headed out of the reserve. I smiled to think of the motorists passing them on the motorway and pointing in amazement and shouting, 'Look kids! Giraffes on the move!'

The capture was a success. It's always a little sad to see an animal leave us but we were pleased that we had excess giraffes to share – it meant we were doing our bit to boost giraffe numbers and diversity in the wild.

'Goodbye guys! Go forth and multiply!' we called, waving them on their way.

31

Humans vs Animals

The list of endangered animals in South Africa makes for distressing reading. We have a job to do to save the rare and precious pangolins, those scale-covered prehistoric creatures with their long noses and tails. Their clever trick of rolling up into an armour-plated ball isn't enough to protect them from heartless humans who poach them for their scales, which, like horns and tusks, have been attributed magical properties. The oribi antelope has been poached almost to extinction. The African wild dog, or painted dog, with its splotchy coat of many colours, is killed as a perceived threat to livestock, and is now one of our most endangered mammals. Seabirds, Cape vultures, snakes, frogs, and even the little Knysna seahorse are under threat, as are many others in South Africa and the world over.

As human populations grow, and habitat declines and fragments, animals and people come into conflict. Understandably, no one wants a huge elephant stomping through their fields of wheat or snacking on their sugarcane or walking across the motorway. But we need to find ways to accommodate our needs and theirs, and protect animals and their habitats.

In 2020 the authorities in China were faced with an

unprecedented and mysterious situation. A herd of fifteen endangered elephants left a nature reserve in Yunnan, lumbering across the country for over a year. No one knows why they went on the move, or where they were going. Were they looking for food? Were they lost? Did the matriarch lead them astray? It is a mystery, but the fact is, they went, walking almost five hundred kilometres.

They ate millions of dollars' worth of crops, and damaged shops and houses, slurped up tonnes of water. But they also charmed the nation. I was very impressed by the way the Chinese responded to these wandering giants. The authorities monitored the elephants twenty-four hours a day, using trackers on foot, and drones which provided extraordinary photographs of these beautiful animals. The pictures captured the Chinese public's attention and went viral across the world.

Officials tried to redirect the elephants to safety and away from residential areas using food and by blocking some roads with trucks. When that wasn't possible, people were temporarily evacuated and relocated so as to avoid conflict between our species and theirs. As mysteriously as they set out, they eventually turned for home. In December 2021, they arrived back at the reserve they had come from. China is one of the few places in the world where the elephant population is growing thanks to extensive conservation efforts, and a hard crackdown on poaching.

Humans have taken over areas where animals live, and animals do what they can to survive. Baboons aren't giving up their turf so easily – if you leave a door open in some areas of the Western Cape, a baboon might pop in and help himself to your fruit bowl, and then mess in your house on the way out! As a result, baboons are now widely seen as a 'problem' species.

We have two troops of baboons at Thula Thula, with about forty individuals in all. Fortunately, they have not been habituated to humans, or learned to associate us with food, so they aren't a problem. I have to admit that I have my own reservations about the baboons. My pack of naughty dogs likes to bark at them and stir up trouble, and I always worry that one day a baboon will have enough of their yapping and come for a fight! Baboons are big and strong, with powerful jaws. My spoiled princes and princesses would be no match for them.

One warm summer afternoon I pushed the dogs off my lap and got up to open the cottage pane glass door to the garden, to let in some cool fresh air. As the door swung open, I came face to face with a huge baboon. He was standing right in front of me, on the doorstep, with his hand stretched out as if he had been about to open the door himself at that same moment. We were about the same size, this big baboon and I, standing eyeball to eyeball at my front door. We both got the fright of our lives. I leapt back with a scream. At the exact moment so did the baboon, barking and baring his huge sharp teeth! Our combined shrieking woke the dogs, who had been enjoying their usual afternoon nap. They emerged from their deep sleep and started barking frantically just as I shut the door on my surprise visitor. The baboon turned and ran off across the lawn, no doubt traumatized by his encounter with a screaming French lady and a pack of wild dogs.

I wondered what would have happened if he'd decided to come in instead. I'm sure that my dogs would have gone for the massive beast – they seem completely unaware of their relative size, and some of the small ones have even been known to try their luck with an elephant. This huge baboon could have torn apart my whole four-legged family. Thank goodness it didn't come to that.

Despite my terrifying baboon encounter, I understand that the baboons are just being baboons! If you watch a troop, you will see that, in fact, baboons are very like us. Very human, especially in their interaction as a family. When a mother baboon strokes her baby's face and looks lovingly into its eyes, you know that they feel the same emotions that we do. They can be quite amusing, too. One troop tends to move through Tented Camp in the late afternoon. The big family tent, number six, has a plate glass door, and the baboons will often stop to examine their reflections. Some catch a terrible fright at the sight of what they think is another baboon and run away. Others will pull faces, or wave their arms, and look quite surprised to see the other baboon respond immediately with the same gesture.

Monkeys face the same problem as baboons do with human interactions – we have taken over what used to be their habitat, cut down trees and built houses on the land, and don't want those monkeys around. I have had recent first-hand experience of the conflict between monkeys and humans. I have an apartment in the seaside town of Umdloti, on the Dolphin Coast, near Durban. What was a peaceful coastal village has become a popular holiday destination, and new residential developments are popping up all around it. Twenty years ago, it was sugarcane fields and natural tropical forest as far as the eye could see – now it's fresh earth and bulldozers.

The cute little vervet monkeys are often spotted around the town. I love watching the little ones playing around their mothers or clinging onto their tummies to catch a ride. A monkey mum and baby make my heart melt. The monkeys have been in the area for centuries, long before we human intruders decided to make our homes there. Our human encroachment has led to habitat loss. They do what they need to survive – and that means stealing food, even raiding houses and apartments like gangs of little thieves.

One weekend in the summer of 2021, Clément left the apartment to come to Thula, and forgot to close the kitchen window. Well, those monkeys spotted it in a flash, and went in to help themselves. Our neighbours didn't have a key to open the flat, and they watched helplessly as the monkeys had a big party inside. The monkeys saw the humans at the window and wouldn't come out. More monkeys arrived, wanting to join the party.

Clément rushed back when we heard about this dire situation. What a vision of horror he returned to! The monkeys had had a fun fight with the eggs, eaten whatever they could get their little hands on, tossing aside what they didn't fancy. And let's just say, they were not the best guests. They certainly did not tidy up after themselves. Poor Clément. A serious clean-up was needed! But you can't blame the monkeys – this is what happens when we humans take the territory of the animals.

We have monkeys at Thula Thula. They really are very cute to watch, gambolling on the lawns of the Lodge in their family groups. They are very mischievous, though. They have learned that the little paper sachets on the tea trays contain sugar, and a bold one will take the opportunity to grab a sachet and toss the whole thing into its mouth. Nor are they above helping themselves to the fresh fruit salad if they think they can get away with it.

Pilfering the sugar is one thing, but stealing expensive electronics is a different matter altogether. Alexandre, a French guest with very little English, came out of his tent at Tented Camp spluttering and gabbling in French. Christiaan could see he was very upset, but couldn't make out what was wrong. When language failed, the man pointed up to a fever tree, where a little grey vervet monkey sat, inspecting his nice new iPhone. The guest had left it temptingly on the veranda and the monkey had picked it up out of curiosity – perhaps he wondered what the humans were doing with those shining things all the time. When he had fiddled a bit and discovered it was not edible, he soon dropped the phone – undamaged, thank goodness. When I visited Tented Camp later, he showed me the monkey's bite marks on the side of the iPhone – a little souvenir to remind him of his time in the bush.

Tracy Rowles, known locally as 'the Monkey Lady', and her team at Umsizi Umkomaas Vervet Rescue save and rehabilitate vervet monkeys who have been injured or orphaned or are found in built-up areas. Their goal is always to release them in the wild. I was so impressed by the work this organization was doing that I offered to give one of the monkey troops a home in the northern part of the Thula Thula reserve.

Tracy had apparently applied for a permit to operate her facility as a rehab centre some time previously and was waiting for a response – a situation that sounded rather familiar to me! The centre had been operating without a legal permit and was given twenty-one days' notice to remove the monkeys. She didn't and was charged and fined. The premises were raided and the animals captured and confiscated – and then culled. The authorities had removed and euthanized fifty-nine monkeys. I could not believe my ears when I heard the horrific news. When I managed to get hold of poor Tracy, she was distraught. She told me that she couldn't sleep at night for fear of another raid, terror that the remaining monkeys – monkeys that she had saved and looked after with so much dedication and love – would be taken and killed.

The incident made the papers, and a top investigative television programme, *Carte Blanche*. The authorities defended their actions, quoting the regulations, and maintaining that the animals were held illegally, but there was quite an outcry from the public, who were horrified and bewildered by what seemed a cruel and unnecessary response. Tracy is busy relocating her sanctuary to a new location, further from town and residential areas, where she and her monkeys will have peace and safety. The final awful irony was that the troop that was slaughtered was the troop that would have come to live here at Thula Thula. If only they had come to us a few weeks earlier.

32

Never a Dull Moment in the Bush

'Boni, quick. Fetch the milk!'

The housekeeper came running, bottle in hand. No, this was not a tea emergency, but a snake drama. Gin had had an encounter with a *mfezi*, the Mozambican spitting cobra. When cornered, this snake rears up, spreads its hood and spits venom with such force that it can travel two or three metres. If the venom lands on the skin or fur it does no harm, but if it gets in the eyes it can cause stinging and blindness. When the rangers come across a *mfezi*, the first thing they do is reach for their sunglasses to protect their eyes. Once they had to wrangle a big fellow who had found his way into my office, and I felt like I was watching a scene from the movie *Men in Black* – four guys wearing their dark shades indoors.

The dogs are not quick learners when it comes to snakes. They just can't resist messing with them. Except for my Gypsy, who is a sensible lap dog, every one of them has got an eye full of cobra venom. Today it was Gin's turn. Fortunately, milk neutralizes the poison. I grabbed the milk from Boni and poured it straight from the bottle into Gin's eyes. Lynda held the squirming dog firmly as the liquid streamed over his face.

'There you go, my silly boy,' I said. 'I hope you've learnt your lesson.'

In response, he gave a good shake, showering me and Lynda and Boni in milk, and ran inside for a restorative nap. He would wake with puffy, swollen eyes, and a vague smell of sour milk, but no permanent damage.

Life in the bush is heaven for dogs, but it is also full of dangers. Peggy the bull terrier was taken by a croc. And every night I hear the hyenas call, I channel my inner hyena whisperer, murmuring a little request to them to stay away from my dogs. But snakes are the biggest risk. Tonic, Gin's brother, was an ill-mannered chap, very naughty and highly energetic. This little Jack Russell had a heart as big as his legs were short. I once saw him climb a tree to get to the monkeys! Tonic's adventurous spirit was the death of him – he met his match when he went down to the river and tried to attack a python. The python came out on top.

When Kim got Zara, she decided that the dog needed snake training. She called on Jack, the snake-crazy manager of the volunteer academy.

'Jack, I need you to bring a non-venomous snake to the office.'

This is the kind of request Jack lives for! He came over immediately with a green spotted bush snake. They are very common here, and inadvertently cause a lot of trouble. They are not venomous, but people who don't know snakes mistake them for green mambas – very poisonous snakes – and freak out!

Kim and Jack introduced Zara to the snake, letting her

examine it and sniff it, but Kim disciplined her against touching or pouncing on it. Her hope was that Zara would remember the lesson and not tangle with a snake if she saw one in the wild.

'Snakes won't bother you if you don't bother them,' Jack was always telling us. 'When did you last hear about someone killed by a snake? As a male human, I'm the most dangerous creature you'll ever encounter. And here you are, having a conversation with me.'

Probably true, statistically, but try telling that to the kitchen staff when they phone in a panic saying: '*Woza!* Come! Help! There's a spitting cobra behind the fridge!'

The rangers are adamant that snakes don't want to get into a fight with dogs or humans – it's just that we stumble into each other from time to time. To avoid this potentially fatal state of affairs, Zulu women used to make ankle bracelets strung with cocoons with little stones in them. Each step creates a rattling vibration and warns the snakes to get out of the way.

As winter approaches, I open my clothes cupboards with trepidation, because cupboards are the perfect place for a snake looking for a warm, sheltered spot to hibernate in. I've tried many alleged snake deterrent methods, sprinkling various special mixtures of lavender, lemon oil, vinegar and who knows what about the place, to no avail. Someone told me that snakes don't like the smell of naphthalene. I figured it can't do any harm to toss a few moth balls about the place – at least our jerseys will be safe from moths, and maybe we'll be safe from snakes, too. I bought a big stash of them and asked the housekeepers at the Lodge and Tented Camp to put them around the place, in corners and under furniture and inside cupboards. I have to say, it seemed to work. We made it through winter without any slithery visitors. It was better for the snakes too. They belong in the bush, not in human houses where they risk being killed or injured.

The way I see it, if it's winter and the snakes are hibernating, then they shouldn't be lying around the garden at night, right? Wrong! One warmish winter evening, I was irritated by the unusually insistent barking of the dogs outside in the garden. I opened the door to quiet them and saw a large blob of something on the ground, surrounded by all the dogs. I got my torch and put on my glasses and saw – to my horror! – that the big dark mass, just metres away from me, was a huge coiled puff adder. That's one of the most dangerous snakes around. And this was a biggie! I believe that they are also deaf, which was lucky for him and for the dogs, with all that mad barking going on. If that puff adder had decided to strike, it would have been the end for whichever dog was in the firing line.

I screamed hysterically for the dogs, calling them into the house, and called for a ranger to come and identify the snake. Khaya was our hero that evening. He confirmed that it was a puff adder and used the snake tongs to delicately lift up the poor terrified creature and take it away from the main house. The snake was no doubt delighted to get away from my yapping little brats and get on with its peaceful existence in the wild. And we were delighted that he was gone.

We try to teach all our dogs not to bark at or chase animals, but they go mad when the buck come into the garden. They might like to chase the nyala for a bit of fun, but the dogs instinctively understand the power of the elephants. When they come to the house, I just have to shhhhh the dogs and they freeze. They just watch those majestic giants walking silently past, following them with their eyes without a sound, not one tail moving.

Lynda's poodles, newly arrived from Pretoria, had no idea about elephant etiquette. The first time they saw the enormous beasts, they launched themselves at the fence, barking hysterically, determined to protect us all from the giants. The whole herd was agitated by the yapping, snarling dogs. ET charged forward letting out an angry, ear-splitting trumpeting sound, and I really thought this was the day that she would plough through the fence to shut them up. An angry elephant – especially one with a few supporters – could easily push over the fence.

Lynda managed to wrangle her dogs into the house, and the elephants settled down. My heart was racing after the stress and the narrow escape. The poodles have never quite got the hang of elephant encounters, so the first thing we do when the elephants come to visit is lock those nervy big city dogs safely inside.

One thing I can say for certain about life in the bush is that you never get a boring day.

33

Why (or How) Did the Elephants Cross the Road?

In August 2021, a team from the wildlife authorities arrived to inspect our fences, and to look at the underpass linking Thula Thula with our new partners, Zulweni. If all went well, we could incorporate the two reserves and we would have almost the requisite 5,000 hectares. And our animals would be able to meet and greet and mingle and – hopefully – mate.

The inspection went well. The only concern was the road between the two reserves. The big question was, will the elephants use the underpass? Never having tried to entice an elephant into an underpass, none of us could say exactly how they would respond. Would they just wander in, cool as you please? Would we have to entice them with tasty lucerne and horse pellets? Would they flat out refuse to go in? Would they come round in their own good time? I couldn't tell you.

But I can tell you – as I told the authorities – that elephants are the most intelligent and resourceful beasts, and if they want to go to the other side of the road, and they have to go through the underpass to do it, that's what they'll do. I've witnessed myself what determined creatures they are. Within twenty-four hours of our original herd arriving at Thula, Nana pushed a tree against the electric fence and the escape artists disappeared.

I've seen a video of an elephant crawling under a fence at Addo Elephant Park. Crawling! Who would imagine it is even possible? I've seen a similar remarkable video from the Nagarhole National Park in India – an elephant climbing over a sturdy wooden barricade designed to keep elephants out of human spaces. He hoisted his substantial bulk over the fence with an agility and determination I could hardly believe. Sadly, this elephant's escape was likely motivated by encroachment of humans on forest land – he was looking for food in the surrounding coffee estates, as there was less available in the forest environment. In this respect, elephants in India suffer many of the same challenges as our wildlife – the loss of land and their ancient migratory routes.

Whatever the reason, these elephant escapes are testament to how intelligent and strategic these special souls can be. That's the sort of attitude we have come to expect from these big fellows – if they want to, they can. I had no doubt they could use the underpass if they wanted to.

In the bush, we are constantly making decisions for which there is no direct precedent and no scientific research. In these matters, I rely very heavily on the people I know and trust, our rangers, our conservation people, the wildlife vets. They work with the animals and the land every day, and have done for decades. Even if they don't know everything, their instincts are good. They certainly know more than me.

The authorities weren't convinced about the underpass and

asked if it would be possible to make a plan for the elephants to cross the road instead.

'Alright, if they can't go under, they will have to go over,' said Michael. 'We'll make a plan.'

With Vusi and Christiaan, he went back to the drawing board to work on the idea of an overpass. Michael had once worked for the road department, so he has good contacts there, and got their input. Within a week we had a plan – an overpass, speed bumps to slow the cars on the road, a cattle grid to keep the elephants off the road, a new fence-line with reinforced fencing around the overpass. And we had the go-ahead from the roads department. Voilà! The new map was drawn and sent to the authorities.

Well, my team and I, we were like the elephant crawling under that fence at Addo – for each objection or rejection, we found a way under or around it. Our determination to keep all our elephant family together will never waver. Setbacks just force us to be more creative and work harder and smarter.

My life sometimes felt like one long wait for permits and permissions. But we had achieved miracles with the new land, and things seemed to be moving forward in a positive direction. I was optimistic that once we joined with Zulweni, at the next permit committee meeting we might get the permit for Thabo's big brother, and the permit for the two male cheetahs, and the permit to reopen our rehab centre which had been sitting empty and unused for two years. Fingers crossed!

My sense of optimism was soon to be dashed.

The consultant we employed to compile our elephant management plan explained exactly how the land/elephant ratio works, and, tapping away at his calculator, confirmed what the authorities had said: we had too many elephants for the size of our reserve. Even at 5,000 hectares, the area was too small. Yes indeed, that's what his calculator said, but I still believed that a decision couldn't be made without looking at the land, and at the happy elephant family that had been living there for twenty-two years. The consultant went away to work on a report and recommendations. When it arrived in my inbox, I couldn't believe what I was reading. He laid out our options – to meet the authorities' requirements we would have to relocate or, failing that, cull some of our elephants. Those were our choices.

Michael, Christiaan and Lynda were in my office when the report came through, and saw my shocked response.

'*Merde, mais c'est pas possible*, I can't believe this!'

'What is it? You are as white as a sheet,' said Lynda. I turned my laptop to face her. I watched her face fall as she read the words on the screen.

'Relocate some of them? Culling . . . ? But what . . . Where . . . We can't . . .'

'They are a family,' I said. 'Imagine if I told you we were going to take one of your children and send them off to live with strangers in Cape Town.'

'I know, it's unthinkable,' she replied.

'Who do they think I must send away? Must I send little Lolo, Nana's lastborn, named for Lawrence after his death? Such a calm, gentle boy. Maybe they would like me to get rid of him. Or maybe they'd like a bigger elephant like Mandla? He's the biggest now, but he first came to visit me under the watchful eye of his mother Nana when he was just a baby. Should he go and live somewhere else?'

Michael swallowed hard, and made as if to answer, but I was in full steam and in no mood to be calmed.

'Or Mabona! She is a lovely, playful girl, I'm sure someone would like to have her. And surely Tom would not mind losing her sister just so we meet the magical number of elephants that some bureaucrat in an office believes is right. But someone else will have to tell Mabona, the Lodge manager, that her namesake is going, because it's not going to be me.'

My anger turned to desperation and tears filled my eyes. I've known some of these elephants for twenty-two years. Some I've known since the day they were born. Each and every elephant in the family has a personal history, and we at Thula Thula know their parents, their personalities, their funny little ways and habits. We can identify each one by sight – by their size, a tear to the ear, a scar or a distinctive angle to a tusk. The rangers can judge their moods on any given day. We know and love each and every one. And they know and love each other.

'We're not sending anyone. We can't,' Christiaan said in horror. 'We have to find a way . . .'

'Did you read the report? If we don't manage to relocate them we will have to . . .' I could hardly finish the sentence. 'They will cull them. Cull. Just another word for murder.'

The tears were falling now as I thought about our beloved

elephants hunted down and shot by men in helicopters, our happy herd ripped apart. Mothers losing sons and daughters. Lifelong friends and companions lost. I know that elephants grieve and mourn their dead loved ones, and that they have long memories. Gypsy, always sensitive to my emotions, appeared at my feet. I picked her up and held her little warm body against me.

'How can this be good for conservation?' asked Michael. 'How is this good for the animals? It doesn't make any sense.' I had always known Michael to be very unflustered, but this development seemed to have shaken his calm.

'It doesn't,' Christiaan agreed. 'You can't just move elephants around. It's not just about numbers, it's about the overall well-being of the herd. The trouble is, an elephant herd is a complex social structure. When you mess with the structure, there are serious consequences. Conservationists learnt that lesson after the disastrous experiments in the Kruger National Park in the 1980s.'

'What happened?' I asked. I was living in Paris in the 1980s, so unsurprisingly, I hadn't been following the elephant situation in the Kruger.

'The vegetation and the water resources were taking a lot of strain from the elephants. The conservationists reckoned there were too many of them, and they decided to cull seventeen thousand elephants.'

Lynda let out a small gasp of disbelief.

'Seventeen thousand?' I said, recoiling in shock. '*Quelle horreur!* That's a massacre.'

Christiaan nodded, solemnly. 'Yes. And they completely

underestimated the social impact of the culling. They mostly killed the older elephants. Now those elephants might have sixty or seventy years of learning and teaching, they are indispensable. And they have relationships. Every elephant is part of a family, and you never know the impact of that individual elephant. There might be three or four young elephants that are dependent on that one for guidance. The upshot of it was that they destroyed the social structure and left a lot of youngsters without experienced elders to guide them and keep them in line. Same thing happened when they first brought elephants to the Pilanesberg. They took young-sters – they are easier to transport, and you can put two in the truck instead of one big elephant, so it saves money. Jeez, but those elephants were a menace! They overturned cars, they killed a rhino. With no authority figures, they didn't know how to behave.'

Thinking about our own elephant family and their deep and complex bonds, I could imagine how devastating that tragic culling experiment must have been to the elephants of the Kruger Park. And how desperately those young elephants in the Pilanesberg needed the firm hand and experience of their elders. I had seen first hand how our own herd fell into disarray when Frankie, the matriarch, died. How, without her to make decisions for the family, they floated around in indecision and confusion.

'And you know what? Getting rid of all those elephants didn't even solve the problem of overgrazing and habitat destruction,' Christiaan continued, shaking his head.

'Really?' said Lynda, surprised. And I must say, I was too.

'What worked in the end was getting rid of some of the man-made dams and watering holes. The animals had to move around to the natural water holes and rivers to find water. The animals cover a bigger area and the impact on any particular place and its vegetation is much less. That's what fixed the problem with overgrazing and the destruction of the bush.'

We sat for a moment, quietly contemplating the situation, then Lynda got up and walked over to the kettle – followed, of course, by her two poodles. I thought about how complex the natural world is, and how, in trying to solve one problem, you can create others. Even the so-called experts aren't always right.

Christiaan sighed. 'If they would only consider the bush, the topography of Thula Thula, they would see that the reserve can handle this number of elephants, and more,' he said for about the hundredth time. It was exasperating. It felt as if we were talking into the void, hoping our logic would be heard.

'If they actually saw our elephants, they would know that they are the happiest herd in South Africa,' I said. This for me was the crux of the matter. If our elephants were unhappy, if they were fighting with each other, or attacking humans, if they were destroying the bush or trying to escape, I would understand and try to find a solution to the problem. But there was no problem.

I had lived with this herd for over two decades. Lawrence and I and our team at Thula Thula had fought for those elephants every day. I didn't claim to have all the answers, but one thing I did know – we had to keep our herd together. There could be no relocation, no culling. We needed more land. And to continue with the contraception, at least for now.

34

Code Black

My phone constantly pings and rings with alerts. I am sometimes tempted to leave my phone off for a few days, just for some peace and quiet. But I can't. We have WhatsApp groups for rangers, a group for monitoring the rhinos and elephants so we know their location at any time, not to mention the everyday communication. The most worrying pings and rings are related to security and our anti-poaching efforts.

We have a system of alert codes, which are communicated as necessary, and which everyone knows and obeys. There are four codes, ranging from the least concerning, Code Yellow, which means that unarmed, 'small-time' poachers have breached the fence-line, to Code Black, which means that there is gunfire exchange in progress. In this case, everyone on the property must return to the lodges and not leave until the situation has been cleared. Fortunately, Code Black is a rare occurrence.

Larry, as head of security and anti-poaching efforts, regularly receives tip-offs about poaching threats, some more worrying than others. Information is key in our battle against the people who would kill our wildlife. We are almost always dealing with imperfect information, and a great deal of uncertainty. Attempts on our animals range from locals poaching for the pot, to heavily armed professionals coming for our rhinos. We don't know in advance what we're dealing with, but in the worst case scenario, the results can be deadly. The pandemic only worsened the situation.

One day I was taking a bit of downtime, wandering about my garden with my afternoon cup of coffee. The weeping boer-bean tree was in full bloom. It is one of the first to flower in spring, and after a long hard winter it is a joy to see its deep red flowers which produce so much nectar that it drips out like rain. I stopped to enjoy the enthusiastic gathering of birds, bees and other animals gorging themselves on the sweet liquid, when my phone vibrated in the pocket of my jacket.

I looked down and saw the name of the messenger.

Larry Erasmus.

Much as I value our security chief, any message from him gives me a feeling of trepidation – too often, it signals trouble. This was to be one of those times.

We have a group of 30 hunters and 30+ dogs on the boundary of Dube Ridge attempting to enter Thula. APU in place and monitoring.

Thirty! That was a small army. Before I had a chance to reply, the next message came.

K9 going on standby

The dog unit – K9 – was being deployed. And then:

ZAP Wing on standby

Air support. This was serious.

He sent through three photos in quick succession, each

showing a stretch of perimeter fence, with the hardened wire cut through and folded back.

Massive group. Armed and with dogs. Looks like a coordinated attempt. We called the SAPS for help and I sent another two vehicles from the office.

This turned out to be a group of 'subsistence' poachers, hoping to bring back animals for the pot or to sell. These poachers get bolder and cleverer all the time. For instance, they have recently started to poach from outside the reserve, shooting an antelope through the fence, and dragging it under the fence. They know our APU can only shoot at them if they are on Thula property. This is one of the reasons why we work with the police, share information and call on them for back-up and to arrest and charge the culprits. This time, the presence of the South African Police Service and Larry's team was enough to dissuade the poachers. No one was hurt, human or animal.

The next time, we weren't so lucky.

Larry: *Undercover investigators called me. A group of poachers from Mpumalanga arrived in KZN 3 days ago. Attempted poaching incident in Pongola area. Seems they have their sights on Thula.*

Françoise: *Do I need to be worried? Who are they?*

Larry: *Looks like we are dealing with a different type of poacher. Possible big scale attack/invasion to get to rhino. It might turn into an extremely dangerous situation.*

Françoise: *How reliable is this intel, do you think?*

Larry: *I've been communicating with this guy for years. First time he's given me a serious warning. SAPS Crime*

Intelligence also involved now. Might relocate a full K9 team to TT. Will keep you posted.

The exchange made my heart race. This threat sounded serious. Larry certainly thought so. If the intel was accurate, we might be in for a battle.

Larry brought out the big guns, quite literally. The day after our exchange of messages, I watched nervously from my house as he arrived with a convoy of vehicles, each one bringing heavily armed APU men. They were kitted out in camouflage and bulletproof vests, rifles over their shoulders. Dogs strained at the leash, eager to get on with the job of sniffing out poachers. The deployment of anti-poaching units looked and felt like the deployment of reinforcements ahead of a battle. It was quite an extraordinary sight to witness – even though I wished with all my heart that it wasn't happening.

The vehicles spread out. Under Larry's instruction, they took up positions at strategic points within the game reserve. Guards would then patrol by foot, hoping to ambush the poachers before they got to the rhinos. The whole reserve was now surrounded by the APU vehicles and men. I felt a bit safer, but still I could not sleep. My ears strained for the sound of gunshots, or the sound of messages arriving on my mobile phone. All I heard was the gentle snoring of the dogs on the floor and on my bed, and the usual night sounds of the bush.

Larry called me first thing in the morning to give me a report – two poachers shot and others arrested. It is awful to think of people dying on our beautiful reserve, but the fact is that we did not choose this war. We want nothing more than a peaceful life in the bush, but we have been forced to defend our precious rhinos, and we did so successfully.

Our rhinos were still alive. But the war was far from over. We might have won this battle, but we knew there would be more incidents like this. In the meantime, experience showed that when poachers have been caught or shot on the reserve, we have a peaceful few months without serious attempts on our wildlife. The rhinos would be safe just a little longer.

Our best protective measure is, of course, dehorning. Sissi's little horn was starting to show in photographs. Can you believe that poachers set up fake social media accounts to follow bush-lovers and reserves, to find photos of rhinos and their location, and track them down? I knew they wouldn't hesitate to kill our baby rhino for that inch or two of useless keratin – in fact, little Sissi was the same age as our little rhinos were when they were killed in our orphanage in 2017. These people are ruthless. The horn had to go. And mum Mona was overdue for dehorning, too. We had waited a little longer than usual because she had a baby. There is always a risk to this kind of operation, and if anything happened to Mona, we would have an orphan rhino to look after.

The vet decided to dart them and dehorn them one at a time, starting with Mona. His worry was that if he darted both together, while one was being worked on, the other might stagger away, or fall into a ditch, or otherwise get into trouble.

Vusi and the volunteers located the rhinos quite easily. Mona was darted and her horn removed. Next up, Sissi. Well, she was having none of it. She took off as fast as her little legs would carry her – she was determined to get away from the big noisy bird. She ran and ran, until I feared she would have a heart attack. The helicopter finally tracked her down and darted her, but for some reason, the first dart didn't work. The vet had to try again, darting her a second time. This time it all went according to plan, and her little horn was removed.

Everyone was on alert and watching their backs in case Mona came back to protect her baby as she had done with Lisa – with nearly fatal results. This time, Mona was nowhere to be seen. When Sissi woke up she wandered around looking for her mum, still wobbly from the anaesthetic. Just as we started to get quite worried, they found each other. They nuzzled each other, clearly happy to be reunited.

Guests from France and America watched with us. The dehorning was a powerful emotional experience for them, and very educational. They were amazed at what went into it – the helicopter, the rangers and their vehicles on the ground, the vet and his team, and of course the small private army, machine guns at the ready, to protect the animals and the horns from poachers. They realized the extreme lengths (and costs) we go to to protect our wildlife – and will hopefully spread awareness about conservation in their own communities.

The war against poaching is ongoing. We can never drop our guard or relax our efforts. It is an exhausting way to live, but we have no choice. Animals' lives are at stake.

35

A Crazy Dream Becomes Reality

Tented Camp is a peaceful place, and simple and close to nature. It's the perfect spot to sit quietly in the shade and wait to see if the hornbills will make an appearance, or stretch out on the comfy bed in your luxury accommodation and consider whether to read a book or just listen to the sound of the cicadas in the trees. But on this day in September 2021, the camp was a hive of activity. The kitchen staff were preparing a celebratory feast. The housekeeping staff were dusting and polishing. Siya and Muzi were helping set out the camp chairs. There was an air of excitement about the place. We were about to welcome important visitors – Thomas Cebekhulu, the chairman, and other representatives of the Ubizo Communal Property Association.

'It's really happening,' I said to Christiaan, who was doing his own bit of housekeeping, nipping a few dead flowers from the mass of aloes and succulents he had grown from cuttings, and that were now in glorious full bloom. 'Today, we sign the Memorandum of Understanding for Dube Ridge.'

The gate guard radioed to say that our guests had arrived, and the rangers would bring them up to the camp. Our lawyer,

Kirsten, and Michael, our partner from neighbouring Zulweni, were already there. We had spent the morning in a three-hour Zoom meeting with the wildlife authorities. After months of misunderstandings and miscommunication, we had met and spoken openly and directly. By the end of the meeting, several issues with our permits had been clarified, problems solved and tremendous progress made.

The ladies who work at Tented Camp put down their spatulas and their dusters, and dashed to get changed. They emerged in magnificent traditional outfits, black skirts trimmed with white braiding. Black and white beads adorned their heads, necks and arms. Each pretty lady was wearing a big smile for the big day. I asked all the rangers to be present for this special occasion. It was especially meaningful for Muzi, who was part of the Cebekhulu community. Our partners had been enthralled to discover that one of theirs had been a part of Thula Thula for so long, trained here and become a knowledgeable and valued game ranger.

At the sound of approaching vehicles, the staff gathered at the entrance to welcome our guests. Thomas and his community leaders arrived, all dressed up most elegantly for the event. With them was Clive Kelly, their lawyer, a charming fellow who I've known for years. His family has been in Zululand for over a century and he speaks fluent Zulu and has an impressive knowledge of Zulu culture.

I welcomed them warmly. The atmosphere was one of excitement and joy. We had so much to celebrate with our new partners and friends. But business first! The lawyers presented the printed documents which Thomas signed on behalf of the Ubizo community, as the chairman, and I signed on behalf of Thula Thula.

'A big and important day for all of us today,' announced Thomas.

And it certainly was. We had a signed agreement. Now we could soon start the fencing. Our plan was to drop the boundaries at the newly fenced part of Lavoni and open up the fences with Zulweni on the 26 November – the first part of our expansion. Next step would be to fully fence this community land, once we have the approval of the authorities. By mid-2022, we would have reached 5,500 hectares. At last! We will be over the magical number of 5,000 hectares and have all paperwork in place.

I must say, a few happy tears came to my eyes as I thought of the significance of this agreement, and what it would mean for conservation and the happiness of our elephant family. Thomas expressed his own delight that these magnificent animals would soon roam Cebekhulu land for the first time in centuries. Being French, I could not imagine celebrating without a glass of champagne, so we had put some excellent South African bubbly on ice. Vusi popped the corks and we all drank a toast to our mutual success. Then we tucked into the celebratory meal.

The Zulu maidens in their traditional attire sang most beautifully and performed a celebratory dance routine. Thomas got up and joined in, and so did his fellow community leaders. I looked in great contentment and gratitude at the people gathered together. Our new partners, Thomas and the rest of Ubizo, are wonderful people who take their commitment to their community, and to conservation seriously. There's great trust and mutual respect between us and a shared vision, as there is with Michael. The Thula Thula team, of course – the rangers, the office staff, the hospitality staff – are like family to me. Everyone was very relaxed, very happy. I knew that we would do great things together. It was a truly wonderful day.

This day made me believe in miracles. A couple of months ago I was at rock bottom and the future was bleak. And look at us now! The immense pressure forced us to move out of our comfort zone and think of new ideas and build new partnerships. Through sheer determination, we had done what seemed impossible. The Greater Zululand Wildlife Conservancy was no longer a crazy dream of Lawrence's. It was a reality. In a world where land for conservation is shrinking every day, we are creating bigger safe spaces for our precious animals. All that remains is to fence the new section, and drop the fences between us and Dube Ridge.

In the words of the great Nelson Mandela: 'It always seems impossible until it's done.'

36

Ups and Downs

Dear Françoise
   All my life I have dreamed of owning a game reserve in Africa, but since reading your book I've changed my mind. I would rather come as a guest . . .

I was a bit puzzled when I read this rather odd opening sentence, but then I just laughed! Of all the letters I received when *An Elephant in My Kitchen* came out, this one from an American lady stays in my mind. It perfectly summed up my situation – the dream life of elephants and the bush and African sunsets, coupled with insane stress and tragic losses and disappointments. I thought of my American correspondent in November 2021, when I faced some of the greatest highs and greatest challenges of my two decades in the bush.

November is a wonderful month at Thula Thula. It's early summer, the bush is green and lush, baby animals are starting to appear, and we are all looking forward to welcoming guests for the busy festive season. November 2021 was looking particularly fine after the dreadful Covid year. The Lodge and the Tented Camp were fully booked, and the volunteer academy was populated by a wonderful gang of young people eager to roll up their sleeves and get involved in conservation. The month started with a spate of good news.

Kirsten phoned, bubbling with excitement, 'Françoise, we got it! We got the permit.'

After six months of waiting, we were finally given permission to bring Rambo, Thabo's 'big brother', to Thula Thula. We would have a new male rhino to be a role model for Thabo and – we hoped – be dad to some baby rhinos! The wildlife authorities had also given the green light to reopen the rehabilitation centre which had been standing empty for two years. This was a huge breakthrough, and I was delighted to think of all the animals we could care for and rewild in the facility.

The highlight, though, was when the fences were dropped between Thula Thula, Zulweni Game Reserve, and the land of Lavoni to combine the reserves to create one large one. We added a further 1,500 hectares to Thula Thula that day. We invited our new partners, guests, friends, volunteers, media and the wildlife authorities to witness the occasion and celebrate with us. We watched as the rangers cut the wire, and rolled the stiff fence away, leaving the wooden poles stark against the horizon. Next, the poles would go. We took out a few – the rest would be removed over the following days.

Amidst all the excitement, the photos and the action, I paused to appreciate the enormity of this moment. The expansion plan we had been working on for years, with so many frustrations and setbacks, was coming to fruition. I thought of all it had taken to get to this point. How often I had been knocked down, and somehow managed to get up and fight on. I thought of how proud and happy Lawrence would be to see this massive step towards his dream of creating a huge conservancy to protect and preserve this beautiful Zululand bush and its animals.

Most importantly, it meant our elephants were safe. We had employed a new consultant ecologist, Tony Roberts, to work on behalf of the Greater Zululand Wildlife Conservancy and take care of all our management plans. He was confident that there would be no culling or relocation of our elephants. The relief I felt was enormous, it was as if a great burden had been lifted from my shoulders.

'Come on Françoise, one more pic of you and Michael,' called Kim, shaking me out of my musings. I complied, giving her a broad and heartfelt smile for the camera. Then it was back to Tented Camp to celebrate. I thanked everyone who had made this day possible, especially our partners and the Thula Thula team. I acknowledged the wildlife authorities who had put me under the pressure that gave me the push to take this step. It is true that having the threat of relocating our elephants made us work harder and more creatively to get the land we needed.

Steve Jobs said, 'Great things in business are never done by one person, they're done by a team of people,' and I do believe that is true. As I looked around Tented Camp at all these happy people eating and chatting and getting to know each other, I was grateful for every one of them. Even the ones who had made my life so stressful at times!

I returned home feeling a great sense of achievement and

the warm glow of happiness. It had been years since I'd felt so confident about the future. The dogs came running up to greet me, jumping and barking in delight as if I'd been away for weeks.

'Hello Gypsy, yes, come on Tina. Down, Zara, come inside . . .' I said, trailing the pack after me into the house. I made a cup of coffee and sat down to see what had arrived in my inbox while I was out and about. I couldn't believe what I was reading.

Every news site was leading with the breaking story about the 'new South African variant' of Covid. Our scientists had been the first in the world to isolate the genome for a worrying new strain, so new it didn't even have a name. I felt a stirring of panic. And then I saw a headline pop up: 'SA/Europe flights cancelled'. All the good energy of the day drained from me in a second. I was struck by a ghastly feeling of déjà vu. We'd been here before; I knew this story.

I ran across the lawn to the reception office. Swazi looked up at me from her screen, shock on her face.

'Françoise, the guests. They're cancelling,' she said.

A torrent of emails was arriving in her inbox, one after the next, cancelling bookings for the festive season.

'*Mon Dieu*. It is a catastrophe,' I said before I could stop myself.

A lovely couple from the States came into reception, ashen faced: 'We've just seen the news. We need to get out of here and get onto flights. Can you help?'

There followed a whirlwind of activity, all the reception and office staff helping guests and volunteers get out of Thula

Thula and onto flights asap. It was exhausting and stressful and chaotic – a near identical replay of the situation when the first lockdown was announced eighteen months before.

By the time the day was over I was exhausted and filled with fear and anxiety. I had faced so many problems, and things were finally looking up. What a marvellous day it had been! And now this.

I honestly didn't know if I could go through it all again – the cancellations, the lockdowns, the empty rooms, the months of worrying about how I would pay the bills. I tossed and turned that night, running through the possible ramifications of this new shock. How long would border closures be in place? What would South Africa do? Would there be another lockdown? Sleep was impossible with so much in my head. Lucy jumped up onto the bed and lay against me. I relaxed the no-dogs-on-bed rule and pretended I didn't notice. It was nice to have the comfort of their warm bodies at this awful time.

I got up the next morning after what felt like half an hour of sleep and got back to work – helping guests and volunteers change their plans, soothing staff worries, and trying to get on with the daily business of running the reserve. My phone was going mad – questions, queries, cancellations, concern from friends. I barely had time to keep up with them all, but in the afternoon, I opened a message from Michael. There was a picture of a nyala, just a very ordinary trail camera picture of a very common antelope. But it was far from ordinary.

The first visitor uses the tunnel, said the caption.

I was overjoyed! The animals were starting to use the

underpass between Thula Thula and Zulweni. I felt certain that the elephants would find their way there, and their curiosity would entice them to explore the new parts of the reserve.

The photograph was circulated amongst the rangers, too. When Khaya saw the photo he asked, 'Which side did it come from and which way is it going?'

When he heard that it was a Zulweni nyala coming into Thula, he was happy. I suspect he wouldn't like 'our' game to leave home.

'We're one big reserve now,' I reminded him. 'There's no "them and us". We are sharing for the greater good.'

Everyone agreed in principle, although I wondered how they'd feel when their beloved elephants finally went walkabout on the other side! This slightly blurry photo of an ordinary antelope was quite literally a light at the end of the tunnel for me – it lifted my spirits from a very low point to, well, a slightly more positive point!

The next day, President Ramaphosa announced that we would not go into hard lockdown, despite the fast spreading new variant, which was subsequently called Omicron. I believe that it would quite simply be too much for the country to bear. This was at least some good news – we could continue to accommodate local guests.

37

Thabo and Ntombi Take the Test

Dr Trever Viljoen had suggested some months back that we take steps to assess exactly what was going on with Ntombi and Thabo's fertility. Why were they not reproducing? Was there a physical problem? Or was the lack of progress in the mating game due to their having been orphaned and brought up by humans? Was there anything we could do to help them along?

Trever spoke to Dr Morné de la Rey who is a world specialist in assisted reproduction in animals, including in vitro fertilization and artificial insemination. He is also a founder of the non-profit Rhino Repro, which is using state of the art fertility techniques to bring the northern white rhinos back from the brink of extinction. When I heard that he had examined and operated on over two hundred and twenty rhinos, I thought he sounded like someone I would like to talk to.

On 8 December, Morné arrived to examine Ntombi and Thabo, and hopefully shed some light on their mysterious – and unsuccessful – sex lives. If necessary, he would operate to correct a problem. Wildlife vet Hendrik Hansen and our vet Trever were there to assist with the procedures and the anaesthesia. All our rangers attended to help with the animals, and to experience this state-of-the-art veterinary intervention. And Kim documented it all, as usual. It was quite an undertaking.

The animals were darted without incident. Now it was down to business. Morné had an array of equipment – tubes and probes, drips and needles, a padded suitcase of vials and bottles. He and his assistant, Carla, laid everything out, a makeshift operating theatre in the bush.

First up, Ntombi. Using an instrument that he and his team had developed specifically for rhinos, Morné scanned her reproductive organs with ultrasound. He saw that the right ovary was inactive, but the left had two functioning follicles. This meant she was inactive but not infertile.

He inserted a needle for the OPU (Ovum Pick-Up) procedure, aspirating two eggs from the left ovary. While Morné was busy with Ntombi, Thabo was being dehorned. Ntombi's op took a little longer than expected, so Trever gave Thabo two light extra doses of anaesthetic to keep him sleeping until Morné and Hendrik arrived. No one wanted Thabo to wake up!

With Ntombi done, now it was Thabo's turn. After a couple of attempts, Morné managed to insert the semen collection wand into Thabo's rectum in order to stimulate his reproductive organs. He was busy with his examination when I heard him mutter, '*Eish*, he's going to wake up!' And he was! That was quite enough probing, as far as Thabo was concerned. He was leaving the party. Morné and Carla stopped what they were doing as Thabo started to move, and then struggled to his feet. Rangers and vets' assistants were thrown off or scattered. Andrew, who was lying across Thabo's neck, clung on, as if he was riding him. Sheldon, a young game ranger working at the volunteer academy, hung on too.

Still blindfolded and with drips and syringes hanging off him, Thabo stumbled about. He shook off the two men, who fortunately landed unhurt in the bush. Meanwhile, everyone who knew Thabo took off as fast as their legs would carry them. We ran for shelter, hurling ourselves into vehicles. I think one of Vusi's guys actually flew to the house. His feet barely touched the ground, and his shoes were lost in the mayhem. While we behaved as if a grenade had gone off in our midst, Morné and Hendrik were quite calm. They seemed rather baffled by our hasty departure. They didn't know that Thabo was Thabo, not a normal rhino, and they didn't know his temper!

Morné had managed to get the specimens and information he needed. Rangers pulled off the medical paraphernalia and blindfold. Thabo wandered off looking rather dopey, wondering what on earth had just happened. After five hours of operations, we retired to the Lodge to get Morné's feedback.

'I'm sorry to say that the news on Thabo isn't good. Tests showed that he isn't producing semen. It's likely that he is infertile, or at least sub-fertile,' said Morné. He went on to explain that rhinos who had experienced trauma or who had been brought up by humans often fail to reproduce success-fully. 'It's not exactly clear why. It seems to have something to do with the hierarchy in rhino society. It seems the young males need the presence and competition of other dominant males to help them develop hormonally, and to act as role models – Thabo has never had a big bull to show him how to cover a cow.'

It was true that Thabo was skittish and submissive around the females. He didn't seem to know what to do with them. I felt sad, knowing that Thabo was likely psychologically and physically unable to reproduce. He was such a fine specimen of a rhino, and such a special guy. It was a pity not to have little baby Thabos running around Thula Thula.

The news on Ntombi was better, said Morné. It was clear that Ntombi wasn't going through the hormonal cycle of rising and falling progesterone and oestrogen that would bring her into heat and enable her to get pregnant. He explained that removing the dormant, non-viable eggs should prompt her body into the normal cycle, and, hopefully, she would release more eggs.

Rhino reproduction is more complex than you might imagine, and a number of physical, social and emotional factors could contribute to the problem. A drought, or too little food can play a role, although that wasn't the case here – there is plenty for the rhinos to eat. It seemed that Ntombi's issues were more likely to do with the social structure and relationships. The presence of a dominant older female like Mona might help a younger, less experienced female to cycle, but the two rhinos weren't hanging out together. The presence of Thabo, as a submissive male in a 'brotherly' relationship with her, would not prompt her into oestrus, but the presence of a sexually active dominant male (here's looking at you, Rambo!) might do so.

Ntombi was thirteen years old but she should have many more productive years left. This was good news. It was likely that Ntombi's cycle would normalize, and with Rambo arriving a week after the procedure, the timing couldn't be better!

Mona wasn't tested, as she has reproduced successfully twice, and there is no reason to think that she won't do it again. Rambo, of course, was a wild rhino who had mated successfully and fathered a number of calves. We had high hopes that both Ntombi and Mona would have babies with the new male.

I was so grateful to have this kind of state-of-the-art, ground-breaking veterinary intervention, and Morné's exceptional knowledge. As Morné said, they are writing the book as they go along!

Just a week after the medical procedure, we were excitedly awaiting a very special delivery. Rambo was coming at last! I sat in the game viewing vehicle, watching a crane lift a huge reinforced steel crate off the back of a truck, my excitement tinged with nervousness. I had to admit that our efforts to increase the rhino population and to help Thabo grow into a well-adjusted male hadn't been a success. Two years ago we had moved heaven and earth to get him a girlfriend, but he showed no interest in her and his bad boy antics continued. Would this long-awaited big brother do the trick?

The crane lowered the crate, the gears straining and heavy chains clanking. Muffled noises came from inside the crate – the sound of two and half tons of rhino shifting about in a confined space. Two assistants from the wildlife capture company climbed on top of the huge metal crate and opened the door.

'He's coming out,' said Kim excitedly.

Rambo started to back out of the crate, his grey rump emerging first. His butt was massive, his thighs thick and strong.

'Are you sure they didn't send us an elephant?' I joked.

His body followed, his broad shoulders and his thick neck and finally his enormous, swaying head, the circular stump of his newly cut horn showing starkly pale against his grey skin.

'He's a big boy,' I said, in admiration. And he truly was a magnificent sight, a strong seventeen-year-old rhino, in his prime.

Fully out, he turned around, surveying his new home. I like to think he was admiring the grassland, lush and sweet after a wet summer, and the green plains beyond, the streams and puddles glinting in the sun. This was his territory now.

'Welcome home Rambo,' I said, as he sniffed the wind. Could he smell Thabo? I wondered. Or Mona?

Rambo hadn't been tranquillized for the trip from Phinda to Thula Thula, but he seemed absolutely calm. I let out a sigh of relief. A capture and release is risky – anything can happen, as we knew, from vehicle breakdowns to injuries. I feared that if he was frightened or confused, he might become aggressive, and charge the vehicles we watched from, admiring and taking photographs. When you've lived with a rhino like

Thabo, you learn to be careful – our vehicles were always positioned so that we could get away in case of a charge – but Rambo seemed to be at ease. He had a quiet confidence about him. He wasn't worried, he could handle this. He lowered his head and grazed for a minute or two. Then he found an appealing spot and lay down for a rest after his drive.

We had decided not to put Rambo into the *boma* to acclimatize. He was a grown-up rhino, experienced in the ways of the bush. The rangers felt that he would be able to look after himself and adjust to his new environment. Sure enough, he seemed completely at home. After a restorative nap, he wandered off into the bush.

Unlike Mona, who had disappeared into the bush with her baby Lisa, and even hid when she heard vehicles coming, Rambo was quite easy to spot. A few days after his arrival, Muzi, Clément, Christiaan and I found him cooling off in the little Mpisi Dam, on the road to Lavoni. We were on our way to see how the elephant herd was enjoying their new playground, when we came upon him. I felt a prickle of apprehension, but again, he was calm, confident, at home. He ignored our presence and got on with the business of enjoying his mud bath.

'Now that's a wild rhino, a normal rhino,' said Christiaan with satisfaction. 'A totally different animal from a rhino who was brought up with humans.'

It saddened me to think of our Thabo, and how his life had been shaped by the loss of his mother and his kin. No matter how we tried to make up for them, how much love and care we gave, we could never give him the skills and upbringing that a rhino family could give him. I still hoped that the presence of Rambo might change something in Thabo, help him develop the strength and confidence that his big brother had. Time would tell.

38

Rambo, Meet Thabo

Rambo settled in, seemingly without a care in the world. The rangers observed him marking out his territory on the Gwala Gwala plains, not far from main house. He sprayed his dung and urine about, indicating to all and sundry that this was his domain.

I couldn't help but wonder what Thabo thought of the smell of this new male. As the only male, and with no one to tell him otherwise, Thabo had the whole of Thula Thula as his territory. How would he take to having a competitor? Would he share? Would he fight to keep his territory?

The rangers kept a close eye on Rambo. They were on tenterhooks, waiting to see how the two big boys would react to each other, and how Rambo would react to the females. As for me, I dreaded the moment when Rambo and Thabo finally came face to face. I reminded myself that rhinos are not aggressive by nature. For the most part, they are quiet and peaceful herbivores. Rambo was used to the presence of other rhinos, male and female. There was no reason to believe he would want to fight with Thabo. Thabo, on the other hand, had never seen a male rhino in all of his thirteen years. His reaction was more unpredictable.

Early in the morning on Christmas Eve, it happened. I received a message from Victor:

Rambo and Thabo together now.

Wow!! Please tell me what happens. Where are you?

Between Numzani's bone area and the Lodge picnic spot. Towards Nana's dam.

I fetched my binoculars and ran out to the mound by the swimming pool which gives a sweeping view of the plains. It was such a miserable, cold day that even the snakes were hiding. It started to drizzle, but I didn't want to waste time going back for a raincoat. Visibility wasn't good, but I found Victor's vehicle. The rhinos were partially obscured by bush.

What are they doing?

They were just staring at each other. And what would they do next? I was on tenterhooks waiting for news from Victor. The mobile phone network in the wilds of Zululand is unreliable at the best of times, but it always seems to disappear for a bit when I'm eagerly waiting for news! Victor had gone quiet, and my increasingly desperate messages weren't being delivered.

Finally, another message came from Victor.

Thabo is lying down. It looks like he's sleeping!

What's Rambo doing?

Watching him sleep

Really, Thabo!

I almost laughed at our little rascal, when he decided to have a nap in front of this huge new male. I remembered that I'd seen Thabo doing the same with the elephants, lying down in the middle of the road in front of them, blocking their way and ignoring them. The elephants were clearly confused by this behaviour. Was that his strategy? To put them off their stride somehow? Was he doing the same with Rambo, or was it just a submissive gesture?

Minutes passed. I watched Victor in the vehicle and waited eagerly for his messages. I was terrified that the two rhinos were going to fight for dominance, and I could not bear the idea of either of them being hurt. I was almost ready to call our vet, just in case, but that seemed silly.

I couldn't stand the suspense. I messaged Victor.

Is he still sleeping?

No. Standing now, waiting. Rambo is in the bush getting his guns

ha ha . . .

Not real laughter, you understand. My heart was in my mouth!

Are they close?

The space in between them is 5m, if not less.

And you?

I am about 100m away. Don't worry, my car is well positioned if anything happens

Thank you Victor. Please keep watching them and take photos if you can.

A few more minutes, and another message from Victor:

Rambo is walking away.

Who knows, maybe Thabo was smarter than I thought. Maybe his strategy worked out. There was no showdown. Rambo walked calmly over to a thick patch of bush and let out a stream of urine, showing that this was his territory now. Thabo got up and followed him, smelling where the urine had landed. Rambo walked on, and Thabo trailed him all the way to the savanna land opposite the *boma*. I could see them clearly now from my vantage point in the garden of the main house, but still I was mystified as to what was happening between them.

Victor arrived back at main house to show me all the photographs he took. The rangers gathered round excitedly. Everyone wanted to hear the detailed first-hand account of the meeting of the rhinos. Victor went over it again and again, passing the photos around. We looked and talked, turning over each piece of information, examining it for clues as to what their behaviour meant, and what might happen next.

'When Thabo lay down, do you think it was a sign of submission?' said Khaya.

'I don't know, but I remember Thabo doing exactly that, lying down in the middle of the road in front of the elephants, blocking their way but completely ignoring them, not engaging at all?' I said.

'I remember. It confused them. They didn't know what he was up to. Maybe that's his clever strategy.'

'The ways of Thabo are unpredictable and unknowable,' said Victor with a smile.

Khaya persisted. 'And then when he followed Rambo? Does that mean they are friends now?'

'Could be. Or maybe he was just curious. He wants to know who this stranger is.'

I was as mystified as everyone else, but I was pleased to

see that there was no indication of aggression between them. That was something to be grateful for.

I tossed and turned that night, wondering what they were up to and hoping that the peace between them continued. At first light, I was up to check the messages from the rangers. Rambo was where we'd left him, but Thabo had left and gone north, to the volunteer camp. He'd put himself as far away as possible from Rambo.

Once again, all we could do was wait and see. It was clear that no humans can say for certain how an animal will react in a new situation.

But now . . . What was going to happen when Rambo met the girls? We were fairly sure that they would already know of each other's existence but they didn't seem to be making any effort to get together. The females were up north, and Rambo was staying put, establishing his own territory further south. I might not know much about rhino reproduction, but I knew that having the boys and the girls on opposite sides of the reserve wasn't going to work!

Our hope was that when one of the females was ready and receptive to the idea of a mate, her scent would bring him to her. A female will spray her scent around to attract males and ward off females, and will make a whistling vocalization. The day Rambo meets the girls will be a day of great celebration. Another dream come true!

39

The Promised Land

Muzi, Christiaan, Clément and I set off towards Lavoni, now one with Thula Thula, no fence between us. The elephants had moved into the new land, and I was eager to see how they liked it. It was a beautiful day to be out and about in the bush. The morning was still cool, and the breeze was fresh. Good rains ensured that the reserve was looking spectacular – every tree and blade of grass a dizzying green, every animal well-fed and healthy.

December is baby season, and the reserve was full of new life – little nyalas, impala, zebra, wildebeest, up and about within minutes of birth, sticking close to their mums as they explored the world. I asked Muzi to stop a moment so we could admire a brand new zebra, just hours old. 'Look how long its legs are,' Christiaan remarked. 'When they're born their legs are the length of a grown zebra's legs. It's to confuse predators who are low to the ground. They just see legs, and don't know that there are vulnerable babies in the herd.'

The December babies are the best Christmas presents imaginable. The sight of them always fills me with joy. But looking at them, I couldn't help but think of Savannah, who we had hoped would be producing her first litter of cubs this

Christmas. And yet, she was by herself, still without a mate, let alone cubs. The matter of the permits for the males was still outstanding. But that was a problem for another day, I told myself.

'Let's go and find those elephants,' I said to Muzi, eager to get back on the road.

On 26 November 2021, we had dropped the fences between Thula Thula, and Lavoni, with friends, staff and media coming to celebrate the occasion, but the real celebration was when our elephant family found their way to the new land. The rangers had alerted me that it had happened, and we were on our way to see them.

We had no trouble seeing where the elephants had been. The evidence was unmistakable. Muzi drove round a bend in the road and stopped suddenly. He got out of the vehicle to inspect a young willow tree lying across the road, preventing access. It had been pulled from the earth, its roots clearly visible. Torn off branches and scattered leaves had been eaten and then crushed under the feet of excited elephants.

Clément and Christiaan jumped down to help Muzi pull the tree out of the road, and we were on our way. It wasn't long before we came across the scattered remains of a cactus tree, torn up, snacked on and stomped on. We continued on our journey to the river in search of the herd. Even in the rugged game drive vehicle and with Muzi's driving skills, we had our work cut out for us. The road was strewn with evidence of their exuberant exploration of their new backyard, and we had to stop a few more times to clear the way.

I heard the familiar crackling sound of elephants feeding, and turned to see Nandi and ET, gorging themselves on the fresh green trees, tearing and snapping young twigs and stuffing them into their mouths. Baby Tom, the scared baby elephant in my kitchen, ran boldly up to the vehicle, her trunk high. I wondered if she recognized me. We watched for a minute or two, and then continued, tracking the river. We came to a clearing in the thick bush, where the elephants had created a new path for themselves. At the bottom of the muddy path was a wide pool, and there were some of Nana's family, swimming and splashing.

The elephants were as we'd never seen them before. They were like kids in a great big theme park, filled with new experiences, new rides and slides, incredible tastes and smells. They seemed stimulated and joyful. For days, the elephants stayed on Lavoni, exploring their new playground, sampling all its delights. I felt sure it would be weeks before they ventured back to familiar territory, but I didn't mind that they stayed away.

Lavoni was everything we had hoped for and more. It truly was the Promised Land for our elephant family. And there was more to come. In March 2022, we would start fencing the Dube Ridge land. When that was complete, the fence between Thula Thula would be dropped and the elephants would have another new stretch of land to explore.

I could not help remembering Frankie and thinking how much she would have loved this new area and how content she would be seeing her family expressing such happiness and excitement. It is now a year since Frankie died, and the herd is still feeling her absence. We humans seldom visit the site of Frankie's death, because it is so inaccessible and also buffalo territory, but the elephants do visit. Who knows what they feel or remember as they pick up her bones, smell them, and taste them?

Brendon is the one most devastated by her death. He was a real mummy's boy. He stayed with his mother as she got sicker and weaker until she chased him away just days before she died, perhaps knowing that the end was coming and not wanting him to experience that loss. Brendon is still missing and mourning his mother and is acting out in strange ways. Victor witnessed a dramatic confrontation between Brendon and the other elephants at Mkhulu Dam. Brendon was behaving badly towards some of the females – touching them inappropriately, trying to mount them without the usual getting-to-know you flirtation. Generally making a nuisance of himself in a most ungentlemanly manner.

'There's no way he was going to get away with that kind of behaviour in this herd,' Victor said, with some disapproval at the chap's bad manners. 'ET saw what was going on and she put a stop to it. She chased him to the dam and pinned him down in the water. Brendon screamed. I couldn't believe the noise he made. But she kept him there, she wouldn't let him get up or get out of the dam.'

I felt some sympathy for poor Brendon, being on the wrong side of ET. She's a formidable elephant, and not one to back down. 'How did it resolve?' I asked Victor.

'Mabula stepped in, as Brendon's big brother. He went into the dam and played with him, calmed him down. It was as if he was saying, "Come on guy, I've been there, I know what to do, just stand down and behave yourself and this whole thing will go away." It was amazing to see.'

Again I was struck by how incredibly emotionally intelligent our elephants are. Here was Mabula stepping in, guiding the youngster, defusing the situation.

'So it all blew over?' I asked.

'Not quite. When Brendon settled down, he tried to get out of the dam, but ET blocked him. She still wouldn't let him out.'

I shook my head in wonder. That ET really was something.

Victor continued. 'Mabula stood in front of her. He was saying, "OK, that's enough, he's learnt his lesson, we're done here." There was a moment of a standoff, the two of them head to head, and everyone watching. The whole herd wanted to see how this drama was going to play out.'

'And?'

'ET let him go.'

There's been a change in Gobisa, too. Now that he's not at his beloved Frankie's side, Gobisa has started to call on the other females. It's as if he's checking on everyone, making sure they are OK. But he also seems to be showing sexual interest in them, and they in him – a big, older bull is a desirable fellow.

Most intriguing is the shifting dynamic within the herd. Of course, we discuss our very own soap opera endlessly. With her strong personality, and the biggest following of family members, Marula is the established matriarch, but she doesn't yet have Frankie's power and authority. Nana's wisdom and experience are the basis of her power, and she doesn't always approve of Marula's decisions.

'Marula is more adventurous, like Frankie. And she's younger,' said Andrew. 'She might decide to go off on some adventure to the other side of the reserve, and Nana is like, "No thank you, we will make our own plans."'

Her family – Tom, Lolo and the rest – will follow her lead, with Nandi at her side, learning the ropes. Despite her gentle nature, Nana is no push-over. Christiaan was out at Lavoni recently when he came across a big tree branch lying on the road. This wasn't unusual. Christiaan got out of the car and pulled the heavy branch out of the way. Nana was nearby, watching him heaving it to the side, into the bush.

'I came back about ten minutes later, and you won't believe what I saw,' he said, shaking his head. 'Nana was hauling that branch back into the road! And she gave me a look which said, quite clearly, "Don't you mess with my tree, young man. If I put the branch here, I want it here. End of story."'

The herd is not as cohesive under Marula's leadership as it was under her predecessor. Frankie's family sticks with Marula, but Nana's family sometimes splits off for a while, and does their own thing. ET sits somewhere in between – sometimes she's with Marula, sometimes with Nana, and sometimes she goes off with her own family. We expect that these dynamics will continue to evolve for a while. With the new land and more space to move around, we might see a real split, with two or even three separate herds emerging.

40

Love and Connection

'Passion is the secret to persistence. Once you fall in love with a vision, it's not possible to give up.' This is one of my favourite quotes from Robin Sharma, one that rings true for me in my own life.

My Thula Thula journey had been entirely driven by love – first I fell in love with Lawrence, and then the land, and then the elephants. Somehow, the elephants got into my soul, and it became my life's work to see them safe and happy. There was no giving up on that vision, no matter how hard the road was at times.

I think back to the start of it all – when Lawrence brought me to this game reserve on my first trip to South Africa thirty-four years ago. What a thrill it was to see the humble impala and the gracious giraffe for the first time at the little Windy Ridge Game Reserve. Never in my wildest dreams could I have imagined that ten years later we would be the owners of that same reserve, which we decided to call Thula Thula, a haven for all wildlife. Nor could I have anticipated that Lawrence would die so suddenly, leaving me responsible for Thula Thula, its people and its animals.

In the ten years since his death, I've faced extreme

challenges – floods, poaching, loss, administrative struggles, betrayal, and most recently the pandemic. I have been through a rollercoaster of emotions that shook me to my core. At times, I felt I couldn't go on. But I had to. I'd fallen in love with the vision, and with the elephants.

Through it all, it was the elephants that gave me a sense of purpose and direction. Their intuition, resilience and sense of family and community had a profound emotional impact on me. They were my example of a perfect model of society, and of a life well lived. They work in unity, under the leadership of a matriarch, for the greater good. This herd of elephants inspired me and showed me the way to dedicate myself to this new life without Lawrence, to move on and carry on the dream, to lead the team to where we are today.

I didn't know how I would live through Lawrence's death, how I would keep Thula Thula going, or how we would survive through Covid. That in 2022 I would be standing here in this beautiful piece of Zululand, where abundant wildlife roams free and safe, was far from certain. And yet, here I am.

Covid taught us that we can never take anything for granted. We must cherish what we have, even the most ordinary moments, because we never know what tomorrow might bring. I would never want to go through the pandemic years again, but I know that I learnt so much in that time and found resources I didn't know I had. I know that adversity and tragedy have a way of pushing us to do things differently, revealing strengths and talents we might not know we have, opening doors to new opportunities.

The secret to survival is the ability to adapt to a new situation. After all these years, I have realized that change is inevitable and that pressure, challenges and obstacles are a necessity for evolution and growth. It is the way we respond to it which makes the difference, and, to our surprise, sometimes it turns out that the impossible is, in fact, achievable.

From my house, as I write this, I can see the kudu, wildebeest and giraffes grazing on the rich grassland, lush and sweet in late summer. In the distance, the elephants are making their way down the far hill, headed for Mkhulu Dam. They have returned home from their Christmas holidays at Lavoni, gorged on delicious new foliage, and with new memories and experiences. Seeing the joy they experienced in their new land gave me a sense of immense happiness and of accomplishment, but we missed them terribly and were delighted to welcome them back after their three months down south. This beautiful sight makes it all worth it, all the harsh learnings of my new responsibilities for which I was so unprepared.

With my team, I've made great strides towards realizing Lawrence's dream of expanding the reserve to protect more animals, and conserve more land for wildlife.

Just weeks ago, we welcomed a new member of the Thula Thula family when we released Luigi the crocodile – unwanted on the farm where he'd found himself – into the dam. The Centre for the Rehabilitation of Wildlife (CROW) asked if we would take him. My reply was, of course! We only have one croc, Gucci, and plenty of dams and water. Their croc would be better off in a reserve than in a river or farm dam where he might be a danger to humans. Luigi was delighted to be here, hurtling out of his long crate into the dam, while the whole hippo family watched the new arrival to their home.

A few weeks before that, Rambo, our big rhino, had come to live at Thula Thula. Already, he seems quite at home. He has now met Thabo, the elephants and the buffalos, greeting them all with a quiet confidence and a complete lack of aggression. Rambo took his time to call on the girls. A month or two after his arrival he began to move slowly but surely into the north of the reserve, where our ladies reside, until finally he came face to face with Ntombi, who was grazing with Thabo. Rambo introduced himself to the beautiful young lady. In rhino love affairs, this means sniffing each other's unique scent. Romantic, *non*?

Thabo wasn't impressed to see this intruder making the moves on his sweetheart. He looked up with a 'what do you think you're up to?' kind of expression, then lowered his head, and took a run at the big fellow. There was a bit of pushing and shoving between the two suitors – with no injuries, thank goodness – and Rambo emerged the victor, walking off into the sunset with Ntombi.

He soon bid her *au revoir* and carried on in a northerly direction. Within days, he made the acquaintance of Mona, and there was more sniffing and getting-to-know-you flirtation.

With the introductions done, and good relations established, Rambo is waiting for one of the two females to be ready to breed. He will know by their scent, and come calling. And then, I hope, rhino grandchildren. After all, that was his mission!

Our poor Thabo, meanwhile, was quite put out by the whole thing. I was heartbroken to see him wandering alone like a lost soul, since he was dumped by Ntombi. He has taken to visiting us at the main house, sometimes spending whole days with us, as if he's looking to us for comfort.

We have more land, more thriving wildlife, and the newly reopened Zululand Rehabilitation Centre, which is already getting calls from all over the region – an indication of how essential this facility is. The lady who sent us Daisy the meerkat called about another meerkat to be rehabilitated and released back in the Kalahari. I am delighted to be able to help so many little souls live healthy and free.

The predator management plan has been resubmitted, this time for the whole of the Greater Zululand Wildlife Conservancy. The next step, I hope, will be receiving the permits for the two male cheetahs. Frustratingly, the dense vegetation has prevented us from seeing Savannah, but we've found evidence of her passage through the reserve – her tracks, and her kills of small antelopes. Savannah is around, and active – just desperately in need of cheetah company, and a mate. It's my great hope that before 2022 is out, there will be cheetah cubs at Thula Thula, our contribution to the survival of this endangered species. I won't give up on this dream. We cannot wait for the day when we welcome the male cheetahs to the reserve.

There's no certainty in the bush. The only thing I can be sure of is that the journey ahead won't be a smooth or easy ride. There will be challenges and obstacles, but giving up is never an option. When you have a cause and a purpose in life, this is what keeps you going through tough times and gives you the strength to keep fighting and pursue the dream.

Life in the bush has been, and will continue to be, a journey of discovery and survival. On good days, it will feel like a grand adventure. Other days, more of a struggle. But every day is an opportunity to make Thula Thula a better and bigger place for animals and humans.

In February, I had the honour of receiving the *Médaille de Chevalier de l'Ordre National du Mérite* from the French Ambassador in Pretoria, for my contribution towards conservation in South Africa. It is lovely to be recognized, but I never make the mistake of thinking that what I do is unique, or that my achievements are mine alone. I dedicated this medal to my amazing team at Thula Thula, who have been by my side all along this journey and helped me to get Thula Thula to where we are today. Any recognition I receive is also recognition for the passionate conservationists, the unsung heroes, who work so hard to make a better world for animals and humans, and save African wildlife from extinction.

You might be reading this book in a snowy land far away. Or a big, bustling city. Or on a beach somewhere. But wherever you are, I hope that you feel a sense of connection to this special little spot on the earth, down here in the south of Africa, to the wildlife and the people who care for them. For we are all connected, for better or for worse, humans and animals, throughout the world, as we share this same planet.

I hope that my story and the story of the elephants of

Thula Thula has made you fall just a little in love with them, too, and understand the need to protect them. To educate people about conservation and to bring awareness about the vanishing wilderness with the urgent necessity to create more space for all our precious wildlife here in Africa, and in the rest of the world. To fight against poaching. To take better care of our planet and all of its inhabitants.

We must all take on this task with passion and determination, so that future generations will be able to share one of the most profound and beautiful experiences imaginable – to see an elephant, a rhino, a cheetah or a lion roaming free in their natural habitat. In the wilds of Africa.

Acknowledgements

I would like to thank my co-writer Kate Sidley for her dedication to this book and for all the documentation and research she has done to create an informative tool on wildlife conservation, and for telling all our numerous bush-life stories and adventures with great style and humour.

Thank you to my super team of game rangers – Siya, Andrew, Victor, Muzi and Khaya – for their fun bush stories and insights into our special elephants, rhinos and all the Thula Thula wild creatures.

A special thank you to Christiaan for his brilliant knowledge of and passion for wildlife and nature conservation – a true learning experience for Kate and me. Thank you, Christiaan, for your immense patience and willingness to help us to produce this book with all the necessary scientific information.

Thank you to my whole team for their love and loyalty to Thula Thula. I hope one day their children and grandchildren will read this book about their lives with pride for their achievements and I hope they will be an inspiration to many future generations.

Thank you to our talented resident photographer Kim for illustrating this book with the most magnificent photos of our wildlife.

Thank you to the special man in my life, Clément, for his love and constant support, and incredible patience with my crazy lifestyle.

To my friend Bella, a fountain of knowledge and wisdom, always of the greatest help and with the best guidance that I value with utmost gratitude.

To all our friends and great supporters of Thula Thula who have been there for us in the darkest moments to help us keep going though the most challenging times. THANK YOU. We would not have made it without you.

To our unique elephant family – Nana, Nandi, Marula, Mabula, ET, Gobisa, Mandla and all – thank you for being an inspiration and for giving me the strength to never give up and lead the Thula Thula team to where we are today.

To Lawrence and our Matriarch Frankie. Your spirits will live on over Thula Thula forever.

With love to all,

Françoise

Picture credits